Studies in Systems, Decision and Control

Volume 15

Series editor

Janusz Kacprzyk, Polish Academy of Sciences, Warsaw, Poland
e-mail: kacprzyk@ibspan.waw.pl

About this Series

The series "Studies in Systems, Decision and Control" (SSDC) covers both new developments and advances, as well as the state of the art, in the various areas of broadly perceived systems, decision making and control- quickly, up to date and with a high quality. The intent is to cover the theory, applications, and perspectives on the state of the art and future developments relevant to systems, decision making, control, complex processes and related areas, as embedded in the fields of engineering, computer science, physics, economics, social and life sciences, as well as the paradigms and methodologies behind them. The series contains monographs, textbooks, lecture notes and edited volumes in systems, decision making and control spanning the areas of Cyber-Physical Systems, Autonomous Systems, Sensor Networks, Control Systems, Energy Systems, Automotive Systems, Biological Systems, Vehicular Networking and Connected Vehicles, Aerospace Systems, Automation, Manufacturing, Smart Grids, Nonlinear Systems, Power Systems, Robotics, Social Systems, Economic Systems and other. Of particular value to both the contributors and the readership are the short publication timeframe and the world-wide distribution and exposure which enable both a wide and rapid dissemination of research output.

More information about this series at http://www.springer.com/series/13304

Christian Servin · Vladik Kreinovich

Propagation of Interval and Probabilistic Uncertainty in Cyberinfrastructure-Related Data Processing and Data Fusion

 Springer

Christian Servin
Information Technology Department
El Paso Community College
El Paso, Texas
USA

Vladik Kreinovich
Department of Computer Science
University of Texas at El Paso
El Paso, Texas
USA

ISSN 2198-4182   ISSN 2198-4190   (electronic)
Studies in Systems, Decision and Control
ISBN 978-3-319-38587-7   ISBN 978-3-319-12628-9   (eBook)
DOI 10.1007/978-3-319-12628-9

Springer Cham Heidelberg New York Dordrecht London

Printed on acid-free paper

Springer International Publishing AG Switzerland is part of Springer Science+Business Media
(www.springer.com)

Preface

Motivation. Data uncertainty affects the results of data processing. So, it is necessary to find out how the data uncertainty propagates into the uncertainty of the results of data processing. This problem is especially important when cyberinfrastructure enables us to process large amounts of heterogeneous data. In the ideal world, we should have an accurate description of data uncertainty, and well-justified efficient algorithms to propagate this uncertainty. In practice, we are often not yet in this ideal situation: the description of uncertainty is often only approximate, and the algorithms for uncertainty propagation are often not well-justified and not very computationally efficient. It is therefore desirable to handle all these deficiencies.

Structure of the Book. In this book:

- in Chapter 2, we explain in what sense the existing approach to uncertainty – as a combination of random and systematic (interval) components – is only an approximation; a more adequate three-component model (with an additional periodic error component) is described and justified, and the existing uncertainty propagation techniques are extended to this model;
- in Chapter 3, we provide a justification for a practically efficient heuristic technique – namely, for a technique based on fuzzy decision-making; and
- in Chapter 4, we explain how the computational complexity of processing uncertainty can be reduced.

All these methods are based on the idealized assumption that we have a good description of the uncertainty of the original data. In practice, often, we do not have this information, we need to extract it from the data. In Chapter 5, we describe how this uncertainty information can be extracted from the data.

Acknowledgments. Our thanks to Dr. Benjamin C. Flores, Dr. Ming-Ying Leung, Dr. Luc Longpré, Dr. Vanessa Lougheed, Dr. William Robertson, Dr. Scott Starks, Dr. Craig Tweedie, Dr. Aaron Velasco, and Dr. Leticia Velazquez for their help, valuable suggestions, and inspiring ideas.

Thanks as well to all the participants of:

- 13th International Symposium on Scientific Computing, Computer Arithmetic, and Verified Numerical Computations SCAN'2008 (El Paso, Texas, September 29 – October 3, 2008),
- 2008 Annual Fall Meeting of the American Geophysical Union AGU'08 (San Francisco, California, December 15–19, 2008),
- 2009 Annual Conference of the Computing Alliance of Hispanic-Serving Institutions CAHSI'2009 (Mountain View, California, January 15–18, 2009),
- 2012 Annual Conference of the North American Fuzzy Information Processing Society NAFIPS'2012 (Berkeley, California, August 6–8, 2012),
- 15th GAMM-IMACS International Symposium on Scientific Computing, Computer Arithmetic, and Verified Numerical Computation SCAN'2012 (Novosibirsk, Russia, September 23–29, 2012),
- Joint World Congress of the International Fuzzy Systems Association and Annual Conference of the North American Fuzzy Information Processing Society IFSA/NAFIPS'2013 (Edmonton, Canada, June 24–28, 2013), and
- 3rd World Conference on Soft Computing (San Antonio, Texas, December 15–18, 2013)

for valuable discussions.

Grant Support. This work was supported in part:

- by the Louis Strokes Alliance for Minority Participation (LSAMP)/Bridge to the Doctorate, GK-12 fellowship program, and Cyber-ShARE Center of Excellence, programs sponsored by the National Science Foundation under grants HRD-0734825, HRD-1242122, DUE-0926721,
- by Grants 1 T36 GM078000-01 and 1R43TR000173-01 from the National Institutes of Health,
- by a grant N62909-12-1-7039 from the Office of Naval Research, and
- by the European Regional Development Fund in the IT4Innovations Centre of Excellence project (CZ.1.05/1.1.00/02.0070).

El Paso, Texas,   *Christian Servin*
August 2014   *Vladik Kreinovich*

Contents

Chapter 1
Introduction

1.1   Need for Data Processing and Data Fusion

Need for Data Processing. To make decisions, we need to have a good understanding of the corresponding processes. To get this understanding, we must obtain and process data.

In general, for many quantities y, it is not easy (or even impossible) to measure them directly. Instead, we measure related quantities $x_1, \ldots, x_n$, and use the known relation $y = f(x_1, \ldots, x_n)$ between x_i and y, and use the measured values $\widetilde{x}_i$ of the auxiliary quantities x_i to estimate the value of the desired quantity y as $\widetilde{y} \stackrel{\text{def}}{=} f(\widetilde{x}_1, \ldots, \widetilde{x}_n)$. Computation of this estimate is what constitutes *data processing*.

Data Processing Is Especially Important for Cyberinfrastructure-Related Data. Data processing is especially important when cyberinfrastructure – i.e., research environments that support advanced data acquisition, data storage, data management, data integration, data mining, data visualization and other computing and information processing services distributed over the Internet – enables us to process large amounts of heterogeneous data.

Example of Heterogenous Data. In geophysics, there are many sources of data for Earth models:

- first-arrival passive seismic data (from actual earthquakes); see, e.g., [38];
- first-arrival active seismic data (from seismic experiments using man-made sources); see, e.g., [3, 20];
- gravity data; and
- surface waves; see, e.g., [42].

Datasets coming from different sources provide complementary information. For example, different geophysical datasets contain different information on earth structure. In general:

© Springer International Publishing Switzerland 2015
C. Servin & V. Kreinovich, *Propagation of Interval & Probabilistic Uncertainty*,
Studies in Systems, Decision and Control 15, DOI: 10.1007/978-3-319-12628-9_1

- some of the datasets provide better accuracy and/or spatial resolution in some spatial areas and in some depths, while
- other datasets provide a better accuracy and/or spatial resolution in other areas or depths.

For example:

- each measured gravity anomaly at a point is the result of the density distribution over a relatively large region of the earth, so estimates based on gravity measurements have (relatively) low spatial resolution;
- in contrast, each seismic data point (arrival time) comes from a narrow trajectory (ray) a seismic wave travels within the earth, so the spatial resolution corresponding to this data is much higher.

Usually, there are several different geophysical datasets available. At present, each of these datasets is often processed separately, resulting in several different models reflecting different aspects of the studied phenomena. It is therefore desirable to combine data from different datasets.

Need for Data Fusion. Often, to decrease uncertainty, we perform several estimates of the same quantity. In this case, before we start processing data, we need to first fuse data points corresponding to the same quantity.

1.2   Need to Take Uncertainty into Consideration

Need to Take Uncertainty into Account. Data comes from measurements and from experts. In both cases, data comes with uncertainty:

- measurements are never 100% accurate; see, e.g., [58], and
- expert estimates are usually even less accurate than measurements.

In the ideal world, the measurement result $\widetilde{x}$ is exactly equal to the desired value x. In practice, however, there is noise, there are imperfection, there are other factors which influence the measurement result. As a consequence, the measurement result $\widetilde{x}$ is, in general, different from the actual (unknown) value x of the quantity of interest, and the *measurement error* $\Delta x \stackrel{\text{def}}{=} \widetilde{x} - x$ is different from 0.

This uncertainty affects the results of data processing. For example, in meteorological and environmental applications, we measure, at different locations, temperature, humidity, wind speed and direction, flows of carbon dioxide and water between the soil and atmosphere, intensity of the sunlight, reflectivity of the plants, plant surface, etc. Based on these *local* measurement results, we estimate the *regional* characteristics such as the carbon fluxes describing the region as a whole – and then use these estimates for predictions. These predictions range from short-term meteorological predictions of weather to short-term environmental predictions of the distribution and survival of different ecosystems and species to long-term predictions of climate change; see, e.g., [1, 37]. Many of these quantities are difficult to measure accurately: for example, the random effects of turbulence and the resulting rapidly changing wind speeds and directions strongly affect our ability to accurately measure carbon dioxide and water flows; see, e.g., [61]. The resulting measurement inaccuracy is one of the main reasons why it is difficult to forecast meteorological, ecological, and climatological phenomena.

It is therefore desirable to describe how the corresponding measurement uncertainty affects the result of data processing.

Measurement Uncertainty: Reminder. Measurements are never 100% accurate. The result $\tilde{x}$ of a measurement is usually somewhat different from the actual (unknown) value x of the quantity of interest.

Bounds on Measurement Error: Interval Uncertainty. Usually, the manufacturer of the measuring instrument provides us with an upper bound Δ on the absolute value of the measurement error $\Delta x \overset{\text{def}}{=} \tilde{x} - x$: $|\Delta x| \leq \Delta$. Because of this bound, once we know the measurement result $\tilde{x}$, we can conclude that the actual (unknown) value x belongs to the interval $[\tilde{x} - \Delta, \tilde{x} + \Delta]$; see, e.g., [28].

Traditional (Probabilistic) Approach to Measurement Errors. In some situations, in addition to the upper bound Δ on (absolute value of) the measurement error, we also know the probabilities of different values $\Delta x \in [-\Delta, \Delta]$.

Namely, when we repeatedly measure the same quantity by the same measuring instrument, we get, in general, slightly different results. Some of these results are more frequent, some less frequent. For each interval of possible values, we can find the frequency with which the measurement result gets into this interval; at first, some of these frequencies change a lot with each new measurement, but eventually, once we have a large number of measurements, these frequencies stabilize – and become *probabilities* of different values of $\tilde{x}$ and, correspondingly, probabilities of different values of measurement error Δx. In other words, the measurement error becomes a *random variable*.

Independence Assumption: Idea. Usually, it is assumed that random variables corresponding to different measurement errors are statistically independent from each other. Indeed, measurement errors corresponding to different sensors usually come from different factors and are, therefore, largely independent.

Independence Assumption: A Formal Description. In statistics, independence of two events A and B means that the probability of A does not depend on B, i.e., that the conditional probability $P(A \mid B)$ of A under condition B is equal to the unconditional probability $P(A)$ of the event A.

The probability $P(A)$ of the event A can be estimated as the ratio $\dfrac{N(A)}{N}$ of the number of cases $N(A)$ when the event A occurred to the total number N of observed cases. Similarly, the probability $P(B)$ of the event B can be estimated as the ratio $\dfrac{N(B)}{N}$ of the number of cases $N(B)$ when the event A occurred to the total number N of observed cases, and the probability $P(A \& B)$ of both events A and B can be estimated as the ratio $\dfrac{N(A \& B)}{N}$ of the number of cases $N(A \& B)$

when both events A and B occurred to the total number N of observed cases. In contrast, to estimate the conditional probability of A given B, we must only take into account cases when B was observed. As a result, we get an estimate $P(A\,|\,B) \approx \dfrac{N(A\,\&\,B)}{N(B)}$. Since $P(A\,\&\,B) \approx \dfrac{N(A\,\&\,B)}{N}$ and $P(B) \approx \dfrac{N(B)}{N}$, we conclude that $N(A\,\&\,B) \approx P(A\,\&\,B) \cdot N$ and $N(B) \approx P(B) \cdot N$ and therefore, $P(A\,|\,B) \approx \dfrac{P(A\,\&\,B) \cdot N}{P(B) \cdot N} = \dfrac{P(A\,\&\,B)}{P(B)}$, so $P(A\,|\,B) \approx \dfrac{P(A\,\&\,B)}{P(B)}$. The larger the sample, the more accurate are these estimates, so in the limit when N tends to infinity, we get the equality $P(A\,|\,B) = \dfrac{P(A\,\&\,B)}{P(B)}$, i.e., equivalently, $P(A\,\&\,B) = P(A\,|\,B) \cdot P(B)$. For independent events, $P(A\,|\,B) = P(A)$ and thus, $P(A\,\&\,B) = P(A) \cdot P(B)$.

So, under the independence assumption, if we have two different series of measurements, resulting in measurement errors Δx and Δy, then the probability $P(\Delta x \in [\underline{x}, \bar{x}] \,\&\, \Delta y \in [\underline{y}, \bar{y}])$ that Δx is in an interval $[\underline{x}, \bar{x}]$ *and* Δy is in an interval $[\underline{y}, \bar{y}]$ is equal to the product of the two probabilities:

- the probability $P(\Delta x \in [\underline{x}, \bar{x}])$ that Δx is in the interval $[\underline{x}, \bar{x}]$, and
- the probability $P(\Delta y \in [\underline{y}, \bar{y}])$ that Δy is in the interval $[\underline{y}, \bar{y}]$:

$$P(\Delta x \in [\underline{x}, \bar{x}] \,\&\, \Delta y \in [\underline{y}, \bar{y}]) = P(\Delta x \in [\underline{x}, \bar{x}]) \cdot P(\Delta y \in [\underline{y}, \bar{y}]).$$

Systematic and Random Error Components. Usually in metrology, the measurement error is divided into two components (see, e.g., [58]):

- the *systematic* error component, which is defined as the expected value (mean) $E(\Delta x)$ of the measurement errors, and
- the *random* error component which is defined as the difference $\Delta x - E(\Delta x)$ between the measurement error Δx and its systematic component $E(\Delta x)$.

Systematic error component is usually described by the upper bound Δ_s on its absolute value: $|E(\Delta x)| \le \Delta_s$, while the random error is usually described by its mean square value

$$\sigma = \sqrt{E\left[(\Delta x - E(\Delta x))^2\right]}.$$

In statistical terms, $\sigma = \sqrt{V}$ is the *standard deviation* of the random variable Δx, i.e., the square root of the *variance* $V = E\left[(\Delta x - E(\Delta x))^2\right]$.

Practical Meaning of Systematic and Random Components of the Measurement Error. The practical meaning of these components – and the practical difference between them – can be described if, in order to improve measurement accuracy, we repeatedly measure the same quantity several times. Once we have several results $\widetilde{x}^{(1)}, \ldots, \widetilde{x}^{(M)}$ of measuring the same (unknown) quantity x, we can then take the arithmetic average

$$\widetilde{x} = \frac{\widetilde{x}^{(1)} + \ldots + \widetilde{x}^{(M)}}{M}$$

as the new estimate.

One can easily see that the measurement error $\Delta x = \widetilde{x} - x$ corresponding to this new estimate is equal to the average of the measurement errors $\Delta x^{(k)} = \widetilde{x}^{(k)} - x$ corresponding to individual measurements:

$$\Delta x = \frac{\Delta x^{(1)} + \ldots + \Delta x^{(M)}}{M}.$$

What are the systematic and random error components of this estimate? Let us start with the systematic error component, i.e., in mathematical terms, with the mean. It is known that the mean of the sum is equal to the sum of the means, and that when we divide a random variable by a constant, its mean is divided by the same constant. All M measurements are performed by the same measuring instrument with the same systematic error $E\left(\Delta x^{(1)}\right) = \ldots = E\left(\Delta x^{(M)}\right)$. Thus, for the sum $\Delta x^{(1)} + \ldots + \Delta x^{(M)}$, the mean is equal to

$$E\left(\Delta x^{(1)} + \ldots + \Delta x^{(M)}\right) = E\left(\Delta x^{(1)}\right) + \ldots + E\left(\Delta x^{(M)}\right) = M \cdot E\left(\Delta x^{(k)}\right).$$

Therefore, the mean of the ratio Δx (which is obtained by dividing the above sum by M) is M times smaller than the mean of the sum, i.e., equal to $E(\Delta x) = E\left(\Delta x^{(k)}\right)$. In other words, the systematic error component does not decrease if we simply repeat the measurements.

In contrast, the random component decreases, or, to be precise, its standard deviation decreases. Indeed, for independent random variables, the variance of the sum is equal to the sum of the variances, and when we divide a random variable by a constant, the variance is divided by the square of this constant. The variance $V = \sigma^2$ of each random error component is equal to $V^{(1)} = \ldots = V^{(M)}$; thus, the variance of the sum $\Delta x^{(1)} + \ldots + \Delta x^{(M)}$ is equal to the sum of these variances, i.e., to

$$V\left[\Delta x^{(1)} + \ldots + \Delta x^{(M)}\right] = V^{(1)} + \ldots + V^{(M)} = M \cdot \left(\sigma^{(k)}\right)^2.$$

Therefore, the variance of the ratio Δx (which is obtained by dividing the above sum by M) is M^2 times smaller than the variance of the sum, i.e., equal to $\dfrac{\left(\sigma^{(k)}\right)^2}{M}$. So, the standard deviation σ (which is the square root of this variance) is equal to $\dfrac{\sigma^{(k)}}{\sqrt{M}}$. In other words, the more times we repeat the measurement, the smaller the resulting random error.

So, when we repeat the same measurement several times, the random error disappears, and the only remaining error component is the systematic error.

How to Estimate Systematic and Random Error Components. As we have just mentioned, a random error component is characterized by its standard deviation σ, while a systematic error component Δx_s is characterized by the upper bound Δ on its absolute value: $|\Delta x_s| \leq \Delta$.

The standard deviation σ of the measuring instrument can be estimated if we repeatedly measure the same quantity x by this instrument. Then, the desired standard deviation can be estimated as the sample standard deviation of the corresponding measurement results $\widetilde{x}^{(1)}, \ldots, \widetilde{x}^{(M)}$:

$$\sigma \approx \sqrt{\frac{1}{M} \cdot \sum_{k=1}^{M} \left(\widetilde{x}^{(k)} - E\right)^2},$$

where $E = \dfrac{1}{M} \cdot \sum_{k=1}^{M} \widetilde{x}^{(k)}$.

To estimate the systematic error component, it is not enough to have the given measuring instrument, we also need to *calibrate* the measuring instrument, i.e., to measure the same quantity x with an additional much more accurate ("standard") measuring instrument – whose measurement result $\widetilde{x}_s$ is assumed to be very close to the actual value x of the measured quantity. Here, $E \approx E(\widetilde{x})$ and $\widetilde{x}_s \approx x$, so the difference $E - x_s$ is approximately equal to $E(\widetilde{x}) - x = E(\widetilde{x} - x) = E(\Delta x)$. Thus, this difference $E - \widetilde{x}_s$ can be used as a good approximation to the systematic error component.

Uncertainty of Expert Estimates. In many practical situations, the measurement results are not sufficient to make reasonable conclusions. We need to supplement measurement results with the knowledge of experts. The use of expert knowledge in processing data is one of the important aspects of *computational intelligence*.

For example, when a medical doctor makes a diagnosis and/or prescribes medicine, he or she is usually not following an algorithm that inputs the patients stats and outputs the name of the disease and the dosage of the corresponding medicine. If medicine was that straightforward, there would have been no need for skilled medical doctors. A good doctor also uses his/her experience, his/her intuition. Similarly, in environmental research, we *measure* temperature, humidity, etc. However, to make meaningful conclusions, it is necessary to supplement these measurement results with *expert estimates* of, e.g., amount of leaves on the bushes ("low", "medium", "high"), state of the leaves – and many other characteristics which are difficult to measure but which can be easily estimated by an expert.

We have mentioned that in data processing, it is important to take into account the uncertainty of measurement results. Expert estimates are usually even much less accurate than measurement results. So, it is even more important to take into account the uncertainty of expert estimates.

One of the main techniques for describing expert uncertainty is *fuzzy techniques*; see, e.g., [31, 50]. In these techniques, for each imprecise (fuzzy) property P (like "small") and for each object x, we describe the degree $\mu_P(x)$ to which this object x satisfies the property P.

For example, one of the most widely used methods of determining the (fuzzy) degree of belief $\mu_P(x)$ (e.g., that a certain temperature x is low) is to poll several experts and take, as $\mu_P(x)$, the proportion of those who think that x satisfies this property.

1.3 How to Propagate Uncertainty in Data Processing

Need to Propagate Uncertainty. In the previous section, we analyzed how to *describe* the uncertainty related to measurements and/or expert estimates. Some quantities can be indeed directly measured or estimated. However, there are many quantities of interest which cannot be directly measured or estimated.

An example of such a quantity is a carbon flux that describes the exchange of carbon between the soil and the atmosphere; see, e.g., [37]. It is difficult to measure this flux directly. Instead, we measure the humidity, wind and concentration of different gases at different height of a special meteorological tower, and then use the results of these measurements to process the data.

In general, for many quantities y, it is not easy (or even impossible) to measure them directly. Instead, we measure related quantities $x_1, \ldots, x_n$, and use the known relation $y = f(x_1, \ldots, x_n)$ between x_i and y to estimate the desired quantity y.

Since measurements come with uncertainty, the resulting estimate is, in general, somewhat different from the actual value of the desired quantity – even when the relation $y = f(x_1, \ldots, x_n)$ is known exactly. It is therefore desirable to *propagate* this uncertainty, i.e., to find out how accurate is the estimate based on (approximate) measurement results.

Possibility of Linearization. The desired quantity y depends on the values x_i:

$$y = f(x_1, x_2, \ldots, x_n).$$

The function $f(x_1, \ldots, x_n)$ describing this dependence is usually smooth (differentiable).

Instead of the actual values x_i, we only know the measurement results $\tilde{x}_i$, results which differ from the actual values by the corresponding measurement errors Δx_i:

$$\tilde{x}_i = x_i + \Delta x_i.$$

After applying the data processing algorithm f to the measurement results $\tilde{x}_i$, we get the estimate $\tilde{y}$ for the desired quantity y:

$$\tilde{y} = f(\tilde{x}_1, \ldots, \tilde{x}_n).$$

We are interested in estimating the difference

$$\Delta y = \tilde{y} - y = f(\tilde{x}_1, \ldots, \tilde{x}_n) - f(x_1, \ldots, x_n).$$

We know that the actual (unknown) value x_i of each measured quantity is equal to

$$x_i = \tilde{x}_i - \Delta x_i.$$

Thus, the desired difference has the form

$$\Delta y = f(\tilde{x}_1, \ldots, \tilde{x}_n) - f(\tilde{x}_1 - \Delta x_1, \ldots, \tilde{x}_n - \Delta x_n).$$

Our objective is to estimate this difference based on the known information about the measurement errors Δx_i.

Measurement errors are usually relatively small, so terms quadratic and of higher order in terms of Δx_i can be safely ignored.

For example, if the measurement error is 10%, its square is 1% which is much much smaller than 10%. If we measure with a higher accuracy, e.g., of 1%, then the square of this value is 0.01% which is even much more smaller than the error itself.

Thus, we can *linearize* the above formula, i.e., expand the dependence of Δy on Δx_i in Taylor series and keep only linear terms in this expansion. As a result, we arrive at the following formula:

$$\Delta y = \sum_i C_i \cdot \Delta x_i,$$

where C_i denotes the corresponding partial derivative $\dfrac{\partial y}{\partial x_i}$.

Separating Random and Systematic Components. As a result of this linearization, we can consider both components separately. Indeed, we know that each measurement error Δx_i consists of two components: systematic component s_i and random component r_i:

$$\Delta x_i = s_i + r_i.$$

The dependence of Δy on the measurement errors Δx_i is linear. Thus, we can represent Δy as the sum of different components coming from, correspondingly, systematic and random errors:

$$\Delta y = \Delta y_s + \Delta y_r,$$

where

$$\Delta y_s = \sum_i C_i \cdot s_i;$$

$$\Delta y_r = \sum_i C_i \cdot r_i.$$

So, it is indeed sufficient to estimate the effect of both types of measurement error components separately.

Using the Independence Assumption. In the following estimations, we will use the above-described typical assumption: that measurement errors corresponding to different measurement are independent. Because of this assumption, we arrive at the following algorithms for estimating different components of Δy.

Propagating Random Component. Propagating random component is the traditional part of error propagation. A natural way to describe the resulting error Δy_r is to use simulations (i.e., a so-called Monte-Carlo approach).

By definition of the random error component, the values r_i and $r_{i'}$ corresponding to different measurements are independent.

There are often many such values. The value Δy_r is thus a linear combination of a large number of independent random variables. It is known that under reasonable conditions, the probability distribution of such a combination tends to normal; this is what is known as the Central Limit Theorem – one of the main reasons why normal distributions are ubiquitous in nature; see, e.g., [71].

A normal distribution is uniquely determined by its mean and standard deviation. We know that each measurement error r_i has mean 0 and a known standard deviation σ_i. The mean of the linear combination is equal to the linear combination of means. Thus, the mean of Δy_r is 0. The standard deviation can be obtained if we repeatedly simulate random errors and take a standard deviation of the corresponding empirical values $\Delta y_r^{(1)}$, $\Delta y_r^{(2)}$, ... Thus, we arrive at the following algorithm.

Propagating Random Component: Algorithm. The random component Δy_r is normally distributed with zero mean. Its standard deviation can be obtained as follows:

- First, we apply the algorithm f to the measurement results $\widetilde{x}_i$ and get the estimate $\widetilde{y}$.
- Then, for $k = 1, \ldots, N$, we do the following:

  - simulate the random errors $r_i^{(k)}$ as independent random variables (e.g., Gaussian) with 0 mean and standard deviation σ_i;
  - form simulated values $x_i^{(k)} = \widetilde{x}_i - r_i^{(k)}$;
  - substitute the simulated values $x_i^{(k)}$ into the data processing algorithm f and get the result $y^{(k)}$.

- Finally, we estimate the standard deviation σ of the random component Δy_r as

$$\sigma = \sqrt{\frac{1}{N} \cdot \sum_{k=1}^{N} \left(y^{(k)} - \widetilde{y} \right)^2}.$$

Mathematical comment. The proof that this algorithm produces a correct result easily follows from the fact that for simulated values, the difference $y^{(k)} - \widetilde{y}$ has the form $\sum_i C_i \cdot r_i^{(k)}$ and thus, has the exact same distribution as $\Delta y_r = \sum_i C_i \cdot \Delta x_i$; see, e.g., [33].

Metrological comment. In some practical situations, instead of the standard deviations $\sigma_i = \sqrt{E[(\Delta x)^2]}$ that describe the *absolute* accuracy, practitioners often describe *relative* accuracy δ_i such as 5% or 10%. In this case, the standard deviation σ_i can be obtained as $\sigma_i = \delta_i \cdot |\widetilde{x}_i|$, i.e., by multiplying the given value δ_i and the absolute value of the signal $\widetilde{x}_i$.

Propagating Systematic Component. Let us now consider the problem of propagating systematic component. By definition, the systematic component Δy_s of the resulting error Δy is equal to $\Delta y_s = \sum_i C_i \cdot s_i$. For each parameter s_i, we know the bound Δ_{si} on its absolute value, so we know that s_i take values between $-\Delta_{si}$ and Δ_{si}.

Thus, to find the smallest and the largest possible value of Δy_s, we need to find out the smallest and the largest possible value of the sum $\Delta y_s = \sum_i C_i \cdot s_i$ when each s_i takes values between $-\Delta_{si}$ and Δ_{si}. One can easily check that the sum is the largest when each term $C_i \cdot s_i$ is the largest.

Each term is a linear function of s_i. A linear function is increasing or decreasing depending on whether the coefficient C_i is positive or negative.

- When $C_i \geq 0$, the linear function $C_i \cdot s_i$ is increasing and thus, its largest possible value is attained when s_i attains its largest possible value Δ_{si}. Thus, this largest possible value is equal to $C_i \cdot \Delta_{si}$.
- When $C_i \leq 0$, the linear function $C_i \cdot s_i$ is decreasing and thus, its largest possible value is attained when s_i attains its smallest possible value $-\Delta_{si}$. Thus, this largest possible value is equal to $-C_i \cdot \Delta_{si}$.

In both cases, the largest possible value is equal to $|C_i| \cdot \Delta_{si}$ and thus, the largest possible value Δ_s of the sum Δy_s is equal to $\Delta_s \overset{\text{def}}{=} \sum_i |C_i| \cdot \Delta_{si}$. Similarly, one can prove that the smallest possible value of Δy_s is equal to $-\Delta_s$.

Thus, we arrive at the following algorithm for computing the upper bound Δ_s on the systematic component Δy_s.

Propagating Systematic Component: Algorithm. The largest possible value Δ_s of the systematic component Δy_s can be obtained as follows:

- First, we apply the algorithm f to the measurement results $\tilde{x}_i$ and get the estimate $\tilde{y}$.
- Then, we select a small value δ and for each direct measurement i, we do the following:

  - for this measurement i, we take $x_i^{(i)} = \tilde{x}_i + \delta$;
  - for all other measurements $i' \neq i$, we take $x_{i'}^{(i)} = \tilde{x}_i$;
  - substitute the resulting values $x_1^{(i)}, \ldots, x_n^{(i)}$ into the data processing algorithm f and get the result $y^{(i)}$.

- Finally, we estimate the desired bound Δ_s on the systematic component Δy_s as

$$\Delta_s = \sum_i \left| \frac{y^{(i)} - \tilde{y}}{\delta} \right| \cdot \Delta_{si}.$$

Metrological comment. In some practical situations, instead of the *absolute* bound Δ_{si} on the systematic error of the i-th sensor, practitioners often describe *relative* accuracy δ_i such as 5% or 10%. In this case, a reasonable way to describe the absolute bound is to determine it as $\Delta_{si} = \delta_i \cdot |\tilde{x}_i|$.

1.4 How to Propagate Uncertainty in Data Fusion

Need for Data Fusion: Reminder. In many real-life situations, we have several measurements and/or expert estimates $\tilde{x}^{(1)}, \ldots, \tilde{x}^{(n)}$ of the same quantity x.

- These values may come from the actual (direct) measurements of the quantity x.
- Alternatively, these values may come from *indirect* measurements of x, i.e., from different models, in which, based on the corresponding measurement results, the i-th model leads to an estimate $\tilde{x}^{(i)}$ for x.

In such situations, it is desirable to fuse these estimates into a single more accurate estimate for x; see, e.g., [58].

Data Fusion: Case of Probabilistic Uncertainty. Let us start with the case when each estimate $\tilde{x}^{(i)}$ is known with the (traditionally described) probabilistic uncertainty, e.g., when each estimation error $\Delta x^{(i)} \stackrel{\text{def}}{=} \tilde{x}^{(i)} - x$ is normally distributed with 0 mean and known standard deviation $\sigma^{(i)}$, and estimation errors $\Delta x^{(i)}$ corresponding to different models are independent.

Comment. In practice, the estimation errors are indeed often normally distributed. This empirical fact can be justified by the Central Limit Theorem, according to which, under certain reasonable conditions, the joint effect of many relatively small errors is (approximately) normally distributed; see, e.g., [71]. For each model based on measurements of a certain type (e.g., gravity or seismic), not only the resulting error of each measurement comes from many different error sources, but also each estimate comes from several different measurements – thus further increasing the number of different error components contributing to the estimation error.

In this case, the probability density for each estimation error $\Delta x^{(i)}$ has the form

$$\frac{1}{\sqrt{2 \cdot \pi} \cdot \sigma^{(i)}} \cdot \exp\left(-\frac{(\Delta x^{(i)})^2}{2 \cdot (\sigma^{(i)})^2}\right) = \frac{1}{\sqrt{2 \cdot \pi} \cdot \sigma^{(i)}} \cdot \exp\left(-\frac{(\tilde{x}^{(i)} - x)^2}{2 \cdot (\sigma^{(i)})^2}\right),$$

and the probability density $\rho(x)$ corresponding to all n estimates is (due to independence) the product of these densities:

$$\rho(x) = \prod_{i=1}^{n} \frac{1}{\sqrt{2 \cdot \pi} \cdot \sigma^{(i)}} \cdot \exp\left(-\frac{(\tilde{x}^{(i)} - x)^2}{2 \cdot (\sigma^{(i)})^2}\right) =$$

$$\left(\prod_{i=1}^{n} \frac{1}{\sqrt{2 \cdot \pi} \cdot \sigma^{(i)}}\right) \cdot \exp\left(-\sum_{i=1}^{n} \frac{(\tilde{x}^{(i)} - x)^2}{2 \cdot (\sigma^{(i)})^2}\right).$$

As a single estimate x for the desired quantity, it is reasonable to select the value for which this probability (density) $\rho(x)$ is the largest (i.e., to use the *Maximum Likelihood* method). Since $\exp(z)$ is an increasing function, maximizing a function

$A \cdot \exp(-B(x))$ is equivalent to minimizing $B(x)$, so we arrive at the following *Least Squares* approach: find x for which the sum $\sum\limits_{i=1}^{n} \dfrac{(\widetilde{x}^{(i)} - x)^2}{2 \cdot (\sigma^{(i)})^2}$ is the smallest possible.

Differentiating this expression with respect to x and equating the derivative to 0, we conclude that $x = \dfrac{\sum\limits_{i=1}^{n} \widetilde{x}^{(i)} \cdot (\sigma^{(i)})^{-2}}{\sum\limits_{i=1}^{n} (\sigma^{(i)})^{-2}}$. The accuracy of this fused estimate can be described by the standard deviation σ for which $\sigma^{-2} = \sum\limits_{i=1}^{n} (\sigma^{(i)})^{-2}$.

Data Fusion: Case of Interval Uncertainty. In some practical situations, the value x is known with interval uncertainty, i.e., we know the interval

$$\mathbf{x}^{(i)} = [\widetilde{x}^{(i)} - \Delta^{(i)}, \widetilde{x}^{(i)} + \Delta^{(i)}]$$

containing the actual (unknown) value of x. This happens, e.g., when we only know the upper bound $\Delta^{(i)}$ on each estimation error $\Delta x^{(i)}$: $|\Delta x^{(i)}| \leq \Delta^{(i)}$. In this case, from the fact that the estimate is $\widetilde{x}^{(i)}$, we can conclude that $|x - \widetilde{x}^{(i)}| \leq \Delta^{(i)}$, i.e., that $\widetilde{x}^{(i)} - \Delta^{(i)} \leq x \leq \widetilde{x}^{(i)} + \Delta^{(i)}$.

For interval uncertainty, it is easy to fuse several estimates. Based on each estimate $\widetilde{x}^{(i)}$, we know that the actual value x belongs to the interval $\mathbf{x}^{(i)}$. Thus, we know that the (unknown) actual value x belongs to the intersection $\mathbf{x} \overset{\text{def}}{=} \bigcap\limits_{i=1}^{n} \mathbf{x}^{(i)} = [\max(\widetilde{x}^{(i)} - \Delta^{(i)}), \min(\widetilde{x}^{(i)} + \Delta^{(i)})]$ of these intervals.

1.5 Challenges

Propagation of Uncertainty: Challenges. In the ideal world, we should have an accurate description of data uncertainty. Based on this description, we should use well-justified efficient algorithms to propagate this uncertainty. In practice, we are often not yet in this ideal situation:

- the description of uncertainty is often only approximate,
- the algorithms for uncertainty propagation are often not well-justified, and
- the algorithms for uncertainty propagation are often not very computationally efficient.

What We Do in This Book. It is desirable to handle all the above challenges. This is what we do in Chapters 2–4.

In Chapter 2, we deal with the fact that the current description of uncertainty is only approximate. Traditionally, measurement uncertainty is represented as a combination of probabilistic (random) and interval (systematic) uncertainty. In Chapter 2, we explain that this approach to representing uncertainty is only an approximation. We show that a more adequate representation of uncertainty leads to a three-component model, with an additional periodic error component. The existing uncertainty propagation techniques are extended to this model. These results are intended to be applied to environmental studies, especially in the analysis of the corresponding time series. As an auxiliary result, the provided mathematical explanation justifies the heuristic techniques that have been proposed to make a description of uncertainty more adequate.

The need to justify heuristic models is dealt with in Chapter 3. Usually, techniques for processing interval and probabilistic uncertainty are well-justified, but many techniques for processing expert (fuzzy) data do not have such a justification. In Chapter 3, we show how a practically efficient heuristic fuzzy technique for decision making under uncertainty can be formally justified.

Challenges related to computational efficiency of uncertainty processing techniques are described in Chapters 4. One of the main reasons why the corresponding computations often take a lot of time is that we need to process a large amount of data. So, a natural way to speed up data processing is:

- to divide the large amount of data into smaller parts,
- process each smaller part separately, and then
- to combine the results of data processing.

In particular, when we are processing huge amounts of heterogenous data, it makes sense to first process different types of data type-by-type and then to fuse the resulting models. This idea is explored in the first sections of Chapter 4.

Even with this idea in place, even when all reasonable algorithmic speed-up ideas have been implemented, the computation time is often still too long. In such situations, the only remaining way to speed up computations is to use different hardware speed-up ideas. Such ideas range from currently available (like parallelization) to more futuristic ideas like quantum computing. While parallelization has been largely well-researched, the use of future techniques (such as quantum computing) in data processing and uncertainty estimation is still largely an open problem. In the last section of Chapter 4, we show how quantum computing can be used to speed up the corresponding computations.

All these formulations, results, and methods are based on the idealized assumption that we have a good description of the uncertainty of the original data. In practice, often, we do not have this information, we need to extract it from the data. In Chapter 5, we describe how this uncertainty information can be extracted from the data.

Chapter 2
Towards a More Adequate Description of Uncertainty

In this chapter, we deal with the fact that the traditional description of uncertainty is only approximate. Namely, measurement uncertainty is usually represented as a combination of probabilistic (random) and interval (systematic) uncertainty. In this chapter, we explain that this approach to representing uncertainty is only an approximation. We show that a more adequate representation of uncertainty leads to a three-component model, with an additional periodic error component. The existing uncertainty propagation techniques are extended to this model. These results are intended to be applied to environmental studies, especially in the analysis of the corresponding time series.

As an auxiliary result, the provided mathematical explanation justifies the heuristic techniques that have been proposed to make a description of uncertainty more adequate.

The results from this chapter were first published in [70, 65]

Need for Time Series. In many applications areas – e.g., in meteorology, in financial analysis – the value of the important variable (temperature, stock price, etc.) changes with time. In order to adequately predict the corresponding value, we need to analyze the observed *time series* – i.e., results of measurements performed at different moments of time – and to make a prediction based on this analysis; see, e.g., [10, 72].

The Traditional Metrological Approach Does Not Work Well for Time Series. As we have mentioned in Chapter 1, in the traditional approach, we represent the measurement error as the sum of two components:

- a *systematic* component which is *the same* for all measurements, and
- a *random* component which is *independent* for different measurements.

When we process time series, this decomposition is insufficient: e.g., usually, there are strong correlations between measurement errors corresponding to consequent measurements.

To achieve a better representation of measurement errors, researchers in environmental science have proposed a semi-empirical idea of introducing the third component of measurement error: the *seasonal* (*periodic*) component; see, e.g., [48].

© Springer International Publishing Switzerland 2015
C. Servin & V. Kreinovich, *Propagation of Interval & Probabilistic Uncertainty*,
Studies in Systems, Decision and Control 15, DOI: 10.1007/978-3-319-12628-9_2

For example, a seasonal error component can represent errors that only happen in spring (this is where the name of this error component comes from), or errors that only happen at night, etc.

From the purely mathematical viewpoint, we can have periodic error components corresponding to all possible frequencies. However, from the physical viewpoint, it makes sense to concentrate on the components with physically meaningful frequencies – and with frequencies which are multiples of these frequencies, e.g., double or triple the daily or yearly frequencies.

For example, in environmental observations, it makes sense to concentrate on daily and yearly periodic errors. If we are interested in the effect of human activity, then we need to add weekly errors – since human activity periodically changes from weekdays to weekends.

According to [48], the idea of using three components of measurement error works well in many practical situations – which leads to two related challenges:

- A *metrological* challenge: how can we explain this success? What is the foundation of this idea?
- A *computational* challenge: how can we efficiently describe this new error component and how can we efficiently propagate it through computations?

In this chapter, we address both challenges:

- we provide a theoretical justification for the semi-heuristic idea of the third error component, and
- we show a natural way for efficiently describing this error component, and show how to efficiently propagate different error components through computations.

First Result: A Theoretical Explanation of the Three-Component Model of Measurement Error. Our objective is to analyze measurement errors $\Delta x(t)$ corresponding to time series. Namely, we want to represent a generic measurement error as a linear combination of several error components.

This division into components can be described on different levels of granularity. Let us consider the level where the granules are the smallest, i.e., where each granule corresponds to a finite-dimensional linear space, i.e., to the linear space whose elements can be determined by finitely many parameters.

Each component of the measurement error is thus described by a finite-dimensional linear space L, i.e., by the set of all the functions of the type $x(t) = c_1 \cdot x_1(t) + \ldots + c_n \cdot x_n(t)$, where $x_1(t), \ldots, x_n(t)$ are given functions, and $c_1, \ldots, c_n$ are arbitrary constants.

In most applications, observed signals continuously (and even smoothly) depend on time, so we will assume that all the functions $x_i(t)$ are smooth (differentiable).

Also, usually, there is an upper bound on the measurement error, so we will assume that each of the the functions $x_i(t)$ are bounded by a constant.

Finally, for a long series of observations, we can choose a starting point arbitrarily. If instead of the original starting point, we take a starting point which is t_0 seconds earlier, then each moment of time which was originally described as moment t is now described as moment $t + t_0$. Then, for describing measurement errors, instead of the original function $x(t)$, we have a new function $x_{t_0}(t)$ for which $x_{t_0}(t) = x(t + t_0)$. It is reasonable to require that the linear space that describes a component of the measurement error does not change simply because we changed the starting point. Thus, we arrive at the following definitions.

Definition 2.1. *We say that a function $x(t)$ of one variable is* bounded *if there exists a constant M for which $|x(t)| \leq M$ for all t.*

Definition 2.2. *We say that a class F of functions of one variable is* shift-invariant *if for every function $x(t) \in F$ and for every real number t_0, the function $x(t + t_0)$ also belongs to the class F.*

Definition 2.3 *By an* error component *we mean a shift-invariant finite-dimensional linear class of functions*

$$L = \{c_1 \cdot x_1(t) + \ldots + c_n \cdot x_n(t)\},$$

where $x_1(t), \ldots, x_n(t)$ are given bounded smooth functions and c_i are arbitrary numbers.

Theorem 2.1. *Every error component is a linear combination of the functions*

$$x(t) = \sin(\omega \cdot t) \text{ and } x(t) = \cos(\omega \cdot t).$$

Comment. Since components are sines and cosines, the decomposition of an error into components is similar to Fourier transform. For readers' convenience, all the proofs are placed at the end of the corresponding chapter.

Practical Consequence of This Result. What are the practical conclusions of this result? We have concluded that the measurement error $\Delta x(t)$ can be described as a linear combination of sines and cosines corresponding to different frequencies ω.

In practice, depending on the relation between the frequency ω and the frequency f with which we perform measurements, we can distinguish between small, medium, and large frequencies:

- frequencies ω for which $\omega \ll f$ are *small*;
- frequencies ω for which $\omega \gg f$ are *large*, and
- all other frequencies ω are medium.

Let us consider these three types of frequencies one by one.

Case When the Frequency ω Is Low. When the frequency ω is low, the corresponding values $\cos(\omega \cdot t)$ and $\sin(\omega \cdot t)$ practically do not change with time: the change period is much larger than the usual observation period.

Thus, we can identify low-frequency components with *systematic* error component – the error component that practically does not change with time.

Case When the Frequency ω Is Large. When the frequency ω is high, $\omega \gg f$, the phases of the values $\cos(\omega \cdot t_i)$ and $\cos(\omega \cdot t_{i+1})$ (or, alternatively, $\sin(\omega \cdot t_i)$ and

$\sin(\omega \cdot t_{i+1})$) corresponding to the two sequential measurements t_i and t_{i+1} differ so much that for all practical purposes, the resulting values of cosine or sine functions are independent.

Thus, high-frequency components can be identified with *random* error component – the error component for which measurement errors corresponding to different measurements are independent.

Case of Intermediate Frequencies ω. In contrast to the cases of low and high frequencies, where the periodicity of the corresponding cosine and sine functions is difficult to observe, components $\cos(\omega \cdot t)$ and $\sin(\omega \cdot t)$ corresponding to medium frequencies ω are observably periodic.

It is therefore reasonable to identify medium-frequency error components with *seasonal (periodic)* components of the measurement error.

This conclusion explains why, in addition to the original physically meaningful frequencies, it is also reasonable to consider their multiples:

- We know that the corresponding error component is a periodic function of time, with the physically meaningful period T_0.
- It is known that every periodic function can be explained into Fourier series, i.e., represented as a linear combination of sines and cosines with frequencies ω which are multiples of the basic frequency $\omega_0 = \dfrac{2\pi}{T_0}$ corresponding to the period T_0.

Three-Component Model Is Now Justified. Thus, we have indeed provided a justification to the three-component model of measurement error.

How to Estimate the Periodic Error Component. In the above text, we explained that the periodic error component is as fundamental as the more traditional systematic and random error components. It is therefore necessary to extend the usual analysis of error components and their propagation to this new type of measurement errors.

As we have mentioned in Chapter 1, for systematic and random error components we know:

- how to describe reasonable bounds on this error component, and
- how to estimate this error component when we calibrate the measuring instrument.

Specifically, the random error component is characterized by its standard deviation σ, while a systematic error component s is characterized by the upper bound Δ: $|s| \leq \Delta$.

The standard deviation σ of the measuring instrument can be estimated if we repeatedly measure the same quantity x by this instrument. Then, the desired standard deviation can be estimated as the sample standard deviation of the corresponding measurement results $\widetilde{x}^{(1)}, \ldots, \widetilde{x}^{(M)}$:

$$\sigma \approx \sqrt{\frac{1}{M} \cdot \sum_{k=1}^{M} \left(\widetilde{x}^{(k)} - E\right)^2},$$

where $E = \dfrac{1}{M} \cdot \sum\limits_{k=1}^{M} \tilde{x}^{(k)}$.

To estimate the systematic error component, it is not enough to have the given measuring instrument, we also need to *calibrate* the measuring instrument, i.e., to measure the same quantity x with an additional much more accurate ("standard") measuring instrument – whose measurement result $\tilde{x}_s$ is assumed to be very close to the actual value x of the measured quantity. Here, $E \approx E(\tilde{x})$ and $\tilde{x}_s \approx x$, so the difference $E - x_s$ is approximately equal to $E(\tilde{x}) - x = E(\tilde{x} - x) = E(\Delta x)$. Thus, this difference $E - \tilde{x}_s$ can be used as a good approximation to the systematic error component.

Since we want to also take into account the periodic error component, it is desirable to provide answers to the above two questions for the periodic error component as well.

How can we describe reasonable bounds for each part of the periodic error component? For each frequency ω, the corresponding linear combination

$$a_c \cdot \cos(\omega \cdot t) + a_s \cdot \sin(\omega \cdot t)$$

can be equivalently represented as $A \cdot \cos(\omega \cdot t + \varphi)$. This is the form that we will use for describing the periodic error component.

Similarly to the systematic error, for the amplitude A, we will assume that we know the upper bound P: $|A| \leq P$.

For phase φ, it is natural to impose a requirement that the probability distribution of phase be invariant with respect to shift $t \to t + t_0$. In other words, the probability distribution for shifted phase $\varphi(t + t_0)$ should be the same as the probability distribution for the original phase $\varphi(t)$. When time is thus shifted, the phase is also shifted by $\varphi_0 = \omega \cdot t_0$. Thus, the requirement leads to the conclusion that the probability distribution for the phase be shift-invariant, i.e., that the corresponding probability density function $\rho(\varphi)$ is shift-invariant $\rho(\varphi) = \rho(\varphi + \varphi_0)$ for every possible shift φ_0. This means that this probability density function must be constant, i.e., that the phase φ is uniformly distributed on the interval $[0, 2\pi]$.

How to Estimate the Periodic Error Component. How can we estimate the periodic error component when calibrating a measuring instrument? When we compare the results of measuring the time series by our measuring instrument and by a standard measuring instrument, we get a sequence of differences $\tilde{x}(t) - \tilde{x}_s(t)$ that approximates the actual measurement errors $\Delta x(t)$.

Periodic error components are sinusoidal components corresponding to several frequencies. In data processing, there is a known procedure for representing each sequence as a linear combination of sinusoids of different frequency – Fourier transform. To find the periodic components, it is therefore reasonable to perform a Fourier Transform; the amplitudes of the Fourier transform corresponding to physically meaningful frequencies (and their multiples) ω will then serve as estimates for the amplitude of the corresponding periodic measurement error component.

Computing Fourier transform is fast: there is a known Fast Fourier Transform (FFT) algorithm for this computation; see, e.g., [9].

In this process, there is still a computational challenge. Indeed, while the standard measuring instrument is reasonably accurate and its measurement results $\widetilde{x}_s(t)$ provide a good approximation to the actual values $x(t)$, these results are still somewhat different from the actual values $x(t)$. Hence, the observed differences $\widetilde{x}(t) - \widetilde{x}_s(t)$ are only approximately equal to the measurement errors $\Delta x(t) = \widetilde{x}(t) - x(t)$. When we apply FFT in a straightforward way, this approximation error sometimes leads to drastic over-estimation of the results; see, e.g., [14, 39]. Because of this fact, many researchers replaced FFT by much slower – but more accurate – error estimation algorithms.

In [39], it was shown how we can modify the FFT techniques so that we get (almost) exact error bounds while being (almost) as fast as the original FFT. So, to estimate the periodic error component, we need to use thus modified FFT algorithm.

How to Propagate Uncertainty in the Three-Component Model. In Chapter 1, we described how to propagate uncertainty described by the two-component model, in which the measurement uncertainty consists of a systematic and a random components. Let us show how algorithms presented in Chapter 1 can be extended to the three-component model, in which we also have a periodic component of measurement uncertainty.

In practical applications, many inputs to the data processing algorithm come from the same sensor at different moments of time. In other words, as inputs, we have the results $\widetilde{x}_i(t_{ij})$ of measuring the values $x_i(t_{ij})$ by the i-th sensor at the j-th moment of time $t_{ij} = t_0 + j \cdot \Delta t_i$, where t_0 is the starting moment of all the measurements, and Δt_i is the time interval between the two consecutive measurements performed by the i-th sensor.

The desired quantity y depends on all these values:

$$y = f(x_1(t_{11}), x_1(t_{12}), \ldots, x_2(t_{21}), x_2(t_{22}), \ldots, x_n(t_{n1}), x_n(t_{n2}), \ldots).$$

Instead of the actual values $x_i(t_{ij})$, we only know the measurement results $\widetilde{x}_i(t_{ij})$, results which differ from the actual values by the corresponding measurement errors $\Delta x_i(t_{ij})$:

$$\widetilde{x}_i(t_{ij}) = x_i(t_{ij}) + \Delta x_i(t_{ij}).$$

After applying the data processing algorithm f to the measurement results $\widetilde{x}_i(t_{ij})$, we get the estimate $\widetilde{y}$ for the desired quantity y:

$$\widetilde{y} = f(\widetilde{x}_1(t_{11}), \widetilde{x}_1(t_{12}), \ldots, \widetilde{x}_n(t_{n1}), \widetilde{x}_n(t_{n2}), \ldots).$$

We are interested in estimating the difference

$$\Delta y = \widetilde{y} - y = f(\widetilde{x}_1(t_{11}), \widetilde{x}_1(t_{12}), \ldots, \widetilde{x}_n(t_{n1}), \widetilde{x}_n(t_{n2}), \ldots) -$$

$$f(x_1(t_{11}), x_1(t_{12}), \ldots, x_n(t_{n1}), x_n(t_{n2}), \ldots).$$

We know that the actual (unknown) value $x_i(t_{ij})$ of each measured quantity is equal to

$$x_i(t_{ij}) = \widetilde{x}_i(t_{ij}) - \Delta x_i(t_{ij}).$$

Thus, the desired difference has the form

$$\Delta y = f(\tilde{x}_1(t_{11}), \ldots, \tilde{x}_n(t_{n1}), \tilde{x}_n(t_{n2}), \ldots) -$$

$$f(\tilde{x}_1(t_{11}) - \Delta x_1(t_{11}), \ldots, \tilde{x}_n(t_{n1}) - \Delta x_n(t_{n1}), \tilde{x}_n(t_{n2}) - \Delta x_n(t_{n2}), \ldots).$$

Our objective is to estimate this difference based on the known information about the measurement errors $\Delta x_i(t_{ij})$.

Possibility of Linearization. As we have mentioned in Chapter 1, measurement errors are usually relatively small, so terms quadratic and of higher order in terms of $\Delta x_i(t_{ij})$ can be safely ignored.

Thus, we can *linearize* the above formula, i.e., expand the dependence of Δy on $\Delta x_i(t_{ij})$ in Taylor series and keep only linear terms in this expansion. As a result, we arrive at the following formula:

$$\Delta y = \sum_i \sum_j C_{ij} \cdot \Delta x_i(t_{ij}),$$

where C_{ij} denotes the corresponding partial derivative $\dfrac{\partial y}{\partial x_i(t_{ij})}$.

Separating Three Components of the Measurement Error. As a result of linearization, we can consider all three components separately.

Indeed, we know that each measurement errors $\Delta x_i(t_{ij})$ consists of three components: systematic component s_i, random component r_{ij}, and periodic component(s) $A_{\ell i} \cdot \cos(\omega_\ell \cdot t_{ij} + \varphi_{\ell i})$ corresponding to different physically meaningful frequencies (and their multiples) ω_ℓ:

$$\Delta x_i(t_{ij}) = s_i + r_{ij} + \sum_\ell A_{\ell i} \cdot \cos(\omega_\ell \cdot t_{ij} + \varphi_{\ell i}).$$

The dependence of Δy on the measurement errors $\Delta x_i(t_{ij})$ is linear. Thus, we can represent Δy as the sum of different components coming from, correspondingly, systematic, random, and periodic errors:

$$\Delta y = \Delta y_s + \Delta y_r + \sum_\ell \Delta y_{p\ell},$$

where

$$\Delta y_s = \sum_i \sum_j C_{ij} \cdot s_i;$$

$$\Delta y_r = \sum_i \sum_j C_{ij} \cdot r_{ij};$$

$$\Delta y_{p\ell} = \sum_i \sum_j C_{ij} \cdot A_{\ell i} \cdot \cos(\omega_\ell \cdot t_{ij} + \varphi_{\ell i}).$$

So, it is indeed sufficient to estimate the effect of all three types of measurement error components separately.

Independence Assumption. In these estimations, we will make a natural assumption: that measurement errors corresponding to different time series are independent. Indeed, as we have mentioned earlier,

- while measurement errors corresponding to measurement by the same sensor at consecutive moments of time are correlated,
- measurement errors corresponding to different sensors usually come from different factors and are, therefore, largely independent.

Because of this assumption, we arrive at the following algorithms for estimating different components of Δy.

Propagating Random Component: Analysis of the Problem. Propagating random component is the traditional part of error propagation. A natural way to describe the resulting error Δy_r is to use simulations (i.e., a Monte-Carlo approach).

By definition of the random error component, the values r_{ij} and r_{ik} corresponding to measurements by the same i-th sensor at different moments of time t_{ij} and $t_{ij'}$ are independent. We are also assuming that the values r_{ij} and $r_{i'j'}$ corresponding to different sensors are independent. Thus, all the values r_{ij} corresponding to different pairs (i, j) are independent.

There are many such values, since each sensor performs the measurements with a high frequency – e.g., one reading every second or every minute. The value Δy_r is thus a linear combination of a large number of independent random variables. It is known that under reasonable conditions, the probability distribution of such a combination tends to normal; this is what is known as the Central Limit Theorem – one of the main reasons why normal distributions are ubiquitous in nature; see, e.g., [71].

A normal distribution is uniquely determined by its mean and standard deviation. We know that each measurement error r_{ij} has mean 0 and a known standard deviation σ_i corresponding to measurements of the i-th sensor. The mean of the linear combination is equal to the linear combination of means. Thus, the mean of Δy_r is 0. The standard deviation can be obtained if we repeatedly simulate random errors and take a standard deviation of the corresponding empirical values $\Delta y_r^{(1)}$, $\Delta y_r^{(2)}$, ... Thus, we arrive at the following algorithm.

Propagating Random Component: Algorithm. The random component Δy_r is normally distributed with zero mean. Its standard deviation can be obtained as follows:

- First, we apply the algorithm f to the measurement results $\widetilde{x}_i(t_{ij})$ and get the estimate $\widetilde{y}$.
- Then, for $k = 1, \ldots, N$, we do the following:

  - simulate the random errors $r_{ij}^{(k)}$ as independent random variables (e.g., Gaussian) with 0 mean and standard deviation σ_i;
  - form simulated values $x_i^{(k)}(t_{ij}) = \widetilde{x}_i(t_{ij}) - r_{ij}^{(k)}$;
  - substitute the simulated values $x_i^{(k)}(t_{ij})$ into the data processing algorithm f and get the result $y^{(k)}$.

- Finally, we estimate the standard deviation σ of the random component Δy_r as

$$\sigma = \sqrt{\frac{1}{N} \cdot \sum_{k=1}^{N} \left(y^{(k)} - \widehat{y}\right)^2}.$$

Mathematical comment. The proof that this algorithm produces a correct result easily follows from the fact that for simulated values, the difference $y^{(k)} - \widehat{y}$ has the form $\sum_i \sum_j C_{ij} \cdot r_{ij}^{(k)}$ and thus, has the exact same distribution as $\Delta y_r = \sum_i \sum_j C_{ij} \cdot \Delta x_i(t_{ij})$; see, e.g., [33].

Metrological comment. In some practical situations, instead of the standard deviations $\sigma_i = \sqrt{E[(\Delta x)^2]}$ that describe the *absolute* accuracy, practitioners often describe *relative* accuracy δ_i such as 5% or 10%. In this case, the standard deviation σ_i can be obtained as $\sigma_i = \delta_i \cdot m_i$, i.e., by multiplying the given value δ_i and the mean square value of the signal

$$m_i = \sqrt{\frac{1}{T_i} \cdot \sum_j (\widetilde{x}_i(t_{ij}))^2},$$

where T_i is the total number of measurements performed by the i-th sensor.

Propagating Systematic Component: Analysis of the Problem. Let us now consider the problem of propagating systematic component. By definition, the systematic component Δy_s of the resulting error Δy is equal to $\Delta y_s = \sum_i \sum_j C_{ij} \cdot s_i$. If we combine terms corresponding to different j, we conclude that $\Delta y_s = \sum_i K_i \cdot s_i$, where $K_i \stackrel{\text{def}}{=} \sum_j C_{ij}$.

The values K_i can be explicitly described. Namely, one can easily see that if for some small value $\delta > 0$, for this sensor i, we take $\Delta x_i(t_{ij}) = \delta$ for all j, and for all other sensors i', we take $\Delta x_{i'}(t_{i'j}) = 0$, then the resulting increase in y will be exactly equal to $\delta \cdot K_i$.

Once we have determined the coefficients K_i, we need to find out the smallest and the largest possible value of the sum $\Delta y_s = \sum_i K_i \cdot s_i$. Each parameter s_i can take any value between $-\Delta_{si}$ and Δ_{si}, and these parameters are independent. Thus, the sum is the largest when each term $K_i \cdot s_i$ is the largest.

Each term is a linear function of s_i. A linear function is increasing or decreasing depending on whether the coefficient K_i is positive or negative.

- When $K_i \geq 0$, the linear function $K_i \cdot s_i$ is increasing and thus, its largest possible value is attained when s_i attains its largest possible value Δ_{si}. Thus, this largest possible value is equal to $K_i \cdot \Delta_{si}$.
- When $K_i \leq 0$, the linear function $K_i \cdot s_i$ is decreasing and thus, its largest possible value is attained when s_i attains its smallest possible value $-\Delta_{si}$. Thus, this largest possible value is equal to $-K_i \cdot \Delta_{si}$.

In both cases, the largest possible value is equal to $|K_i| \cdot \Delta_{si}$ and thus, the largest possible value Δ_s of the sum Δy_s is equal to $\Delta_s \stackrel{\text{def}}{=} \sum_i |K_i| \cdot \Delta_{si}$. Similarly, one can prove that the smallest possible value of Δy_s is equal to $-\Delta_s$.

Thus, we arrive at the following algorithm for computing the upper bound Δ_s on the systematic component Δy_s.

Propagating Systematic Component: Algorithm. The largest possible value Δ_s of the systematic component Δy_s can be obtained as follows:

- First, we apply the algorithm f to the measurement results $\tilde{x}_i(t_{ij})$ and get the estimate $\tilde{y}$.
- Then, we select a small value δ and for each sensor i, we do the following:

    - for this sensor i, we take $x_i^{(i)}(t_{ij}) = \tilde{x}_i(t_{ij}) + \delta$ for all moments j;
    - for all other sensors $i' \neq i$, we take $x_{i'}^{(i)}(t_{i'j}) = \tilde{x}_i(t_{i'j})$;
    - substitute the resulting values $x_{i'}^{(i)}(t_{i'j})$ into the data processing algorithm f and get the result $y^{(i)}$.

- Finally, we estimate the desired bound Δ_s on the systematic component Δy_s as

$$\Delta_s = \sum_i \left| \frac{y^{(i)} - \tilde{y}}{\delta} \right| \cdot \Delta_{si}.$$

Metrological comment. In some practical situations, instead of the *absolute* bound Δ_{si} on the systematic error of the i-th sensor, practitioners often describe *relative* accuracy δ_i such as 5% or 10%. In this case, a reasonable way to describe the absolute bound is to determine it as $\Delta_{si} = \delta_i \cdot a_i$, i.e., by multiplying the given value δ_i and the mean absolute value of the signal

$$a_i = \frac{1}{T_i} \cdot \sum_j |\tilde{x}_i(t_{ij})|.$$

Numerical Example. Let us consider a simple case when we are estimating the difference between the average temperatures at two nearby locations. For example, we may be estimating the effect of a tree canopy on soil temperature, by comparing the temperature at a forest location with the temperature at a nearby clearance location. Alternatively, we can be estimating the effect of elevation of the temperature by comparing the temperatures at different elevations. In this case, we use the same frequency $\Delta t_1 = \Delta t_2$ for both sensors, so $t_{1j} = t_{2j} = t_j$. The difference in average temperatures is defined as

$$y = f(x_1(t_0), x_2(t_0), x_1(t_1), \ldots, x_2(t_1), \ldots, x_1(t_n), x_2(t_n)) =$$

$$\frac{x_1(t_0) + \ldots + x_1(t_n)}{n+1} - \frac{x_2(t_0) + \ldots + x_2(t_n)}{n+1}.$$

Let us assume that the know upper bound on the systematic error of the first sensor is $\Delta_{s1} = 0.1$, and the upper bound on the systematic error of the second sensor is $\Delta_{s2} = 0.2$. We perform measurements at three moments of time $t = 0, 1, 2$. During these three moments of time, the first sensor measured temperatures $\tilde{x}_1(t_0) = 20.0$, $\tilde{x}_1(t_1) = 21.9$, and $\tilde{x}_1(t_2) = 18.7$, and the second second measured temperatures $\tilde{x}_2(t_0) = 22.4$, $\tilde{x}_2(t_1) = 23.5$, and $\tilde{x}_2(t_2) = 21.0$. In this case, the estimate $\tilde{y}$ for the desired difference y between average temperatures is equal to

$$\tilde{y} = \frac{20.0 + 21.9 + 18.7}{3} - \frac{22.4 + 23.5 + 21.0}{3} = 20.2 - 22.3 = -2.1.$$

According to our algorithm, we first select a small value δ, e.g., $\delta = 0.1$.

Then, we modify the results of the first sensor while keeping the results of the second sensor unchanged. As a result, we get $x_1^{(1)}(t_0) = \tilde{x}_1(t_0) + \delta = 20.0 + 0.1 = 20.1$, and similarly $x_1^{(1)}(t_1) = 22.0$ and $x_1^{(1)}(t_2) = 18.8$; we also get $x_2^{(1)}(t_0) = \tilde{x}_2(t_0) = 22.4$, and similarly $x_2^{(1)}(t_1) = 23.5$ and $x_2^{(1)}(t_2) = 21.0$. For thus modified values, we get

$$y^{(1)} = \frac{x_1^{(1)}(t_0) + x_1^{(1)}(t_1) + x_1^{(1)}(t_2)}{3} - \frac{x_2^{(1)}(t_0) + x_2^{(1)}(t_1) + x_2^{(1)}(t_2)}{3} =$$

$$\frac{20.1 + 22.0 + 18.8}{3} - \frac{22.3 + 23.5 + 21.0}{3} = 20.3 - 22.3 = -2.0.$$

Similarly, we modify the results of the second sensor while keeping the results of the first sensor unchanged. As a result, we get $x_1^{(2)}(t_0) = \tilde{x}_1(t_0) = 20.0$, and similarly $x_1^{(2)}(t_1) = 21.9$ and $x_1^{(2)}(t_2) = 18.7$; we also get $x_2^{(2)}(t_0) = \tilde{x}_2(t_0) + \delta = 22.4 + 0.1 = 22.5$, and similarly $x_2^{(2)}(t_1) = 23.6$ and $x_2^{(2)}(t_2) = 21.1$. For thus modified values, we get

$$y^{(2)} = \frac{x_1^{(2)}(t_0) + x_1^{(2)}(t_1) + x_1^{(2)}(t_2)}{3} - \frac{x_2^{(2)}(t_0) + x_2^{(2)}(t_1) + x_2^{(2)}(t_2)}{3} =$$

$$\frac{20.0 + 21.9 + 18.7}{3} - \frac{22.4 + 23.6 + 21.1}{3} = 20.2 - 22.4 = -2.2.$$

Finally, we estimate the desired bound Δ_s on the systematic component $\Delta_s y$ as

$$\Delta_s = \frac{|y^{(1)} - \tilde{y}|}{\delta} \cdot \Delta_{s1} + \frac{|y^{(2)} - \tilde{y}|}{\delta} \cdot \Delta_{s2} =$$

$$\frac{|(-2.0) - (-2.1)|}{0.1} \cdot 0.1 + \frac{|(-2.2) - (-2.1)|}{0.1} \cdot 0.3 = 1 \cdot 0.1 + 1 \cdot 0.3 = 0.4.$$

Propagating Periodic Component: Analysis of the Problem. Finally, let us consider the problem of propagating the periodic components. By definition, the periodic-induced component $\Delta y_{p\ell}$ of the resulting error Δy is equal to

$$\Delta y_{p\ell} = \sum_i \sum_j C_{ij} \cdot A_{\ell i} \cdot \cos(\omega_\ell \cdot t_{ij} + \varphi_{\ell i}),$$

i.e., to

$$\Delta y_{p\ell} = \sum_i \sum_j C_{ij} \cdot A_{\ell i} \cdot (\cos(\omega_\ell \cdot t_{ij}) \cdot \cos(\varphi_{\ell i}) - \sin(\omega_\ell \cdot t_{ij}) \cdot \sin(\varphi_{\ell i})).$$

By combining the terms corresponding to different j, we conclude that

$$\Delta y_{p\ell} = \sum_i A_{\ell i} \cdot K_{ci} \cdot \cos(\varphi_{\ell i}) + \sum_i A_{\ell i} \cdot K_{si} \cdot \sin(\varphi_{\ell i}),$$

where $K_{ci} \stackrel{\text{def}}{=} \sum_j C_{ij} \cdot \cos(\omega_\ell \cdot t_{ij})$ and $K_{si} \stackrel{\text{def}}{=} \sum_j C_{ij} \cdot \sin(\omega_\ell \cdot t_{ij})$.

The values K_{ci} and K_{si} can be explicitly described. Namely:

- One can easily see that if for some small value $\delta > 0$, for this sensor i, we take $\Delta x_i(t_{ij}) = \delta \cdot \cos(\omega_\ell \cdot t_{ij})$ for all j, and for all other sensors i', we take $\Delta x_{i'}(t_{i'j}) = 0$, then the resulting increase in y will be exactly equal to $\delta \cdot K_{ci}$.
- Similarly, if for this sensor i, we take $\Delta x_i(t_{ij}) = \delta \cdot \sin(\omega_\ell \cdot t_{ij})$ for all j, and for all other sensors i', we take $\Delta x_{i'}(t_{i'j}) = 0$, then the resulting increase in y will be exactly equal to $\delta \cdot K_{si}$.

Once we have determined the coefficients K_{ci} and K_{si}, we need to describe the probability distribution of the sum $\Delta y_{p\ell} = \sum_i A_{\ell i} \cdot K_{ci} \cdot \cos(\varphi_{\ell i}) + \sum_i A_{\ell i} \cdot K_{si} \cdot \sin(\varphi_{\ell i})$.

We assumed that all φ_i are independent (and uniformly distributed). Thus, for the case of multiple sensors, we can apply the Central Limit Theorem and conclude that the distribution of the sum $\Delta y_{p\ell}$ is close to normal.

In general, normal distribution is uniquely determined by its first two moments: mean and variance (or, equivalently, standard deviation). The mean of each sine and cosine term is 0, so the mean of the sum $\Delta y_{p\ell}$ is zero as well. Since the terms corresponding to different sensors are independent, the variance of the sum is equal to the sum of the variances of individual terms. For each i, the mean of the square

$$(A_{\ell i} \cdot K_{ci} \cdot \cos(\varphi_{\ell i}) + A_{\ell i} \cdot K_{si} \cdot \sin(\varphi_{\ell i}))^2 =$$

$$A_{\ell i}^2 \cdot (K_{ci}^2 \cdot \cos^2(\varphi_{\ell i}) + K_{si}^2 \cdot \sin(\varphi_{\ell i}) + 2 \cdot K_{ci} \cdot K_{si} \cdot \cos(\varphi_{\ell i}) \cdot \sin(\varphi_{\ell i}))$$

is equal to $\dfrac{1}{2} \cdot A_{\ell i}^2 \cdot (K_{ci}^2 + K_{si}^2)$. Thus, the variance of the sum is equal to

$$\frac{1}{2} \cdot \sum_i A_{\ell i}^2 \cdot (K_{ci}^2 + K_{si}^2).$$

Each amplitude $A_{\ell i}$ can take any value from 0 to the known bound $P_{\ell i}$. The above expression monotonically increases with $A_{\ell i}$, and thus, it attains its largest value when $A_{\ell i}$ takes the largest value $P_{\ell i}$. Thus, the largest possible value of the variance is equal to $\dfrac{1}{2} \cdot \sum_i P_{\ell i}^2 \cdot (K_{ci}^2 + K_{si}^2)$.

Thus, we arrive at the following algorithm for computing the upper bound $\sigma_{p\ell}$ of the standard deviation of the periodic-induced component $\Delta y_{p\ell}$ on the approximation error Δy.

Propagating Periodic Component: Algorithm. The upper bound $\sigma_{p\ell}$ on the standard deviation of the periodic-induced component $\Delta y_{p\ell}$ can be obtained as follows:

- First, we apply the algorithm f to the measurement results $\tilde{x}_i(t_{ij})$ and get the estimate $\tilde{y}$.
- Then, we select a small value δ and for each sensor i, we do the following:

  - for this sensor i, we take $x_i^{(ci)}(t_{ij}) = \tilde{x}_i(t_{ij}) + \delta \cdot \cos(\omega_\ell \cdot t_{ij})$ for all moments j;
  - for all other sensors $i' \neq i$, we take $x_{i'}^{(ci)}(t_{i'j}) = \tilde{x}_i(t_{i'j})$;
  - substitute the resulting values $x_{i'}^{(ci)}(t_{i'j})$ into the data processing algorithm f and get the result $y^{(ci)}$;
  - then, for this sensor i, we take $x_i^{(si)}(t_{ij}) = \tilde{x}_i(t_{ij}) + \delta \cdot \sin(\omega_\ell \cdot t_{ij})$ for all moments j;
  - for all other sensors $i' \neq i$, we take $x_{i'}^{(si)}(t_{i'j}) = \tilde{x}_i(t_{i'j})$;
  - substitute the resulting values $x_{i'}^{(si)}(t_{i'j})$ into the data processing algorithm f and get the result $y^{(si)}$.

- Finally, we estimate the desired bound $\sigma_{p\ell}$ as

$$\sigma_{p\ell} = \sqrt{\frac{1}{2} \cdot \sum_i P_{\ell i}^2 \cdot \left(\left(\frac{y^{(ci)} - \tilde{y}}{\delta} \right)^2 + \left(\frac{y^{(si)} - \tilde{y}}{\delta} \right)^2 \right)}.$$

Metrological comment. In some practical situations, instead of the *absolute* bound $P_{\ell i}$ on the amplitude of the corresponding periodic error components, practitioners often describe *relative* accuracy $\delta_{\ell i}$ such as 5% or 10%. In this case, a reasonable way to describe the absolute bound is to determine it as $\sigma_i = \delta_i \cdot m_i$, i.e., by multiplying the given value δ_i and the mean square value of the signal

$$m_i = \sqrt{\frac{1}{T_i} \cdot \sum_j (\tilde{x}_i(t_{ij}))^2}.$$

Example. To test our algorithm, we have applied it to compute the corresponding error component in the problem of estimating carbon and water fluxes described in the paper [48], where such the notion of a periodic error component was first introduced. Our numerical results are comparable with the conclusions of that paper. In the future, we plan to apply all the above algorithms to the results obtained by the sensors on the Jornada Experimental Range Eddy covariance tower and on the nearby robotic tram, and by the affiliated stationary sensors [18, 25, 26, 27, 36, 60].

Conclusion. In many application areas, it is necessary to process time series. In this processing, it is necessary to take into account uncertainty with which we know the corresponding values. Traditionally, measurement uncertainty has been classified into systematic and random components. However, for time series, this classification is often not sufficient, especially in the analysis of seasonal meteorological and environmental time series. To describe real-life measurement uncertainty more accurately, researchers have come up with a semi-empirical idea of introducing a new type of measurement uncertainty – that corresponds to periodic errors. In this chapter, we provide a mathematical justification for this new error component, and describe efficient algorithms for propagating the resulting three-component uncertainty.

Proof of Theorem 2.1

$1°$. Let us first use the assumption that the linear space L is shift-invariant.

For every i from 1 to n, the corresponding function $x_i(t)$ belongs to the linear space L. Since the error component is shift-invariant, we can conclude that for every real number t_0, the function $x_i(t + t_0)$ also belongs to the same linear space. Thus, for every i from 1 to n and for every t_0, there exist values $c_1, \ldots, c_n$ (possibly depending on i and on t_0) for which

$$x_i(t + t_0) = c_{i1}(t_0) \cdot x_1(t) + \ldots + c_{in}(t_0) \cdot x_n(t). \tag{2.1}$$

$2°$. We know that the functions $x_1(t), \ldots, x_n(t)$ are smooth. Let us use the equation (2.1) to prove that the functions $c_{ij}(t_0)$ are also smooth (differentiable).

Indeed, if we substitute n different values $t_1, \ldots, t_n$ into the equation (2.1), then we get a system of n linear equations with n unknowns to determine n values $c_{i1}(t_0)$, $\ldots, c_{in}(t_0)$:

$$x_i(t_1 + t_0) = c_{i1}(t_0) \cdot x_1(t_1) + \ldots + c_{in}(t_0) \cdot x_n(t_1);$$

$$\ldots$$

$$x_i(t_n + t_0) = c_{i1}(t_0) \cdot x_1(t_n) + \ldots + c_{in}(t_0) \cdot x_n(t_n).$$

The solution of a system of linear equations – as determined by the Cramer's rule – is a smooth function of all the coefficients and right-hand sides. Since all the right-hand sides $x_i(t_j + t_0)$ are smooth functions of t_0 and since all the coefficients $x_i(t_j)$ are constants (and thus, are also smooth), we conclude that each dependence $c_{ij}(t_0)$ is indeed smooth.

$3°$. Now that we know that all the functions $x_i(t)$ and $c_{ij}(t_0)$ are differentiable, we can differentiate both sides of the equation (2.1) with respect to t_0 and then take $t_0 = 0$. As a result, we get the following systems of n differential equations with n unknown functions $x_1(t), \ldots, x_n(t)$:

$$\dot{x}_i(t) = c_{i1} \cdot x_1(t) + \ldots + c_{in} \cdot x_n(t),$$

where $\dot{x}_i(t)$ denotes derivative over time, and c_{ij} denoted the value of the corresponding derivative $\dot{c}_{ij}$ when $t_0 = 0$.

3°. We have shown that the functions $x_1(t), \ldots, x_n(t)$ satisfy a system of linear differential equations with constant coefficients.

It is known that a general solution of such system of equations is a linear combination of functions of the type $t^k \cdot \exp(\lambda \cdot t)$, where k is a natural number (non-negative integer), and λ is a complex number. Specifically, λ is an eigenvalue of the matrix c_{ij}, and the value $k > 0$ appears when we have a degenerate eigenvalue, i.e., an eigenvalue for which there are several linearly independent eigenvectors.

4°. Every complex number λ has the form $a + i \cdot \omega$, where a is its real part and ω is its imaginary part. So:

$$\exp(\lambda \cdot t) = \exp(a \cdot t) \cdot \cos(\omega \cdot t) + i \cdot \exp(a \cdot t) \cdot \sin(\omega \cdot t).$$

Thus, every function $x_i(t)$ can be represented as a linear combination of expressions of the types $t^k \cdot \exp(a \cdot t) \cdot \cos(\omega \cdot t)$ and $t^k \cdot \exp(a \cdot t) \cdot \sin(\omega \cdot t)$.

5°. Now, we can use the requirement that the functions $x_i(t)$ are bounded.

5.1°. Because of this requirement, we cannot have $a \neq 0$:

- for $a > 0$, the function is unbounded for $t \to +\infty$, while
- for $a < 0$, the function is unbounded for $t \to -\infty$.

So, we must have $a = 0$.

5.2°. Similarly, if $k > 0$, the corresponding function is unbounded. Thus, we must have $k = 0$.

6°. Thus, every function $x_i(t)$ is a linear combination of the trigonometric functions $x(t) = \sin(\omega \cdot t)$ and $x(t) = \cos(\omega \cdot t)$.
    The theorem is proven.

Chapter 3
Towards Justification of Heuristic Techniques for Processing Uncertainty

As we have mentioned in Chapter 1, some heuristic methods for processing uncertainty lack justification and are, therefore, less reliable. Usually, techniques for processing interval and probabilistic uncertainty are well-justified, but many techniques for processing expert (fuzzy) data still lack such a justification. In this chapter, we show how a practically efficient heuristic fuzzy technique for decision making under uncertainty can be formally justified.

The results from this chapter were first published in [66].

3.1 Formulation of the Problem

Traditional Approach to Decision Making. Traditional decision making techniques (see, e.g., [12, 13, 30, 41, 59]) deal with the problems in which the quality of each possible alternative is characterized by the values of several quantities. For example, when we buy a car, we are interested in its cost, its energy efficiency, its power, size, etc. Specifically, for each of these quantities, we usually have some desirable range of values.

Sometimes, there is only one alternative that satisfies all these requirements. In other real-life situations, there are several different alternatives all of which satisfy all these requirements. In such cases, the traditional decision making approach usually assumes that there is an objective function that describes the user's preferences; the corresponding techniques then enable us to select an alternative with the largest possible value of this objective function.

Traditional Approach to Decision Making: Limitations. The traditional approach to decision making assumes:

- that the user knows exactly what he or she wants – i.e., knows the objective function – and
- that the user also knows exactly what he or she will get as a result of each possible decision.

© Springer International Publishing Switzerland 2015
C. Servin & V. Kreinovich, *Propagation of Interval & Probabilistic Uncertainty*,
Studies in Systems, Decision and Control 15, DOI: 10.1007/978-3-319-12628-9_3

In practice, the user is often uncertain:

- the user is often uncertain about his or her own preferences, and
- the user is often uncertain about possible consequences of different decisions.

It is therefore desirable to take this uncertainty into account when we describe decision making

Fuzzy Target Approach. To describe actual decision making, the authors of [19, 23, 22, 24, 21, 76, 77] proposed an alternative approach. In this approach, to properly take uncertainty into account, for each numerical characteristic of a possible decision, we form two fuzzy sets:

- first, we form a fuzzy set $\mu_i(x)$ describing the users' ideal value;
- then, we form the fuzzy set $\mu_a(x)$ describing the users' impression of the actual value.

For example, a person wants a well done steak, and the steak comes out as medium well done. In this case, we form a fuzzy set $\mu_i(x)$ corresponding to "well done", and we form a fuzzy set $\mu_a(x)$ corresponding to "medium well done".

How can we estimate to what extent the actual result is satisfactory? All we know is the membership functions corresponding to ideal (desired) and actual. If these sets were crisp, then we could say that it is possible that the proposed solution is satisfactory if some of the possibly actual values is also desired. In the fuzzy case, when we only have degrees describing to what exact each value is possible and to what extent each value is desired, we can consider the degree to which the proposed solution can be desired. To find this degree, we can use the fact that a possible decision is satisfactory if:

- either the actual value is x_1, and this value is desired,
- or the actual value is x_2, and this value is desired,
- ...,

where $x_1, x_2, \ldots$, go over all possible values of the desired quantity.

We know the membership functions $\mu_i(x)$ and $\mu_a(x)$. This means that for each value x_k, we know the the degree $\mu_a(x_k)$ with which this value is actual, and the degree $\mu_a i(x_k)$ to which this value is desired. If we use $\min(a,b)$ to describe "and" (the simplest possible choice of an "and"-operation [31, 50]), then we can estimate the degree to which the value x_k is both actual *and* desired as

$$\min(\mu_a(x_k), \mu_i(x_k)).$$

If we now use $\max(a,b)$ to describe "or" (the simplest possible choice of an "or"-operation [31, 50]), then we can estimate the degree d to which the two fuzzy sets match as

$$d = \max_x \min(\mu_a(x), \mu_i(x)).$$

How can we elicit the corresponding membership functions? In principle, membership functions can have different shapes. It is known, however, that in many applications (e.g., in intelligent control), the actual shape of a membership function does not affect the result. Thus, it is reasonable to use the simplest possible membership functions – symmetric triangular ones [31, 50].

To describe a symmetric triangular function, it is sufficient to know the support of this function, i.e., the interval $[\underline{x}, \overline{x}]$ on which this function is different from 0. This interval can also be described as $[\tilde{x} - \Delta_x, \tilde{x} + \Delta_x]$, where:

- $\tilde{x} = \dfrac{\underline{x} + \overline{x}}{2}$ is the interval's midpoint, and

- $\Delta_x = \dfrac{\overline{x} - \underline{x}}{2}$ is the interval's half-width.

The corresponding membership function:

- linearly increases from 0 to 1 on the first half-interval $[\tilde{x} - \Delta_x, \tilde{x}]$, and then
- linearly decreases from 1 to 0 on the second half-interval $[\tilde{x}, \tilde{x} + \Delta_x]$.

As we have just mentioned, once we know the interval, we can uniquely determine the corresponding membership function. So, to elicit the membership function from the user, it is sufficient to elicit the corresponding interval. How can we elicit the interval from the user? To elicit this interval, we can simply ask the users which values are possible, and then take the smallest of these possible values as $\underline{x}$ and the largest of these possible values as $\overline{x}$.

So, to get the membership function $\mu_i(x)$ describing the desired situation, we can ask the user for all the values $a_1, \ldots, a_n$ which, in their opinion, satisfy the requirement, and then take the smallest of these values as $\underline{a}$ and the largest of these values as $\overline{a}$.

Similarly, to get the membership function $\mu_a(x)$ describing the result of a proposed decision, we can ask the user for all the values $b_1, \ldots, b_m$ which, in their opinion, satisfy the corresponding property (like "medium well done"), and then take the smallest of these values as $\underline{b}$ and the largest of these values as $\overline{b}$.

Fuzzy Target Approach: Successes. The above approach works well, e.g., in predicting how the customers buying handcrafted souvenirs select among "almost-desirable" souvenirs when their "ideal" souvenir is not available.

Fuzzy Target Approach: The Remaining Problem. The above method is somewhat heuristic: it is based on selecting the simplest possible membership function and the simplest possible "and"- and "or"-operations. If we use more complex membership functions and "and"- and "or"-operations, we will get different results.

The fact that the existing approach works well in practice indicates that there should be a justification for this method which goes beyond simplicity. Such a justification would give us more confidence in using these techniques.

What We Do in This Chapter. In this chapter, we provide a justification for the above target-based fuzzy decision procedure.

3.2 Solution to the Problem

What We Know: Reminder. As we have mentioned earlier, while the procedure that we want to justify uses fuzzy techniques, all we know is two intervals:

- an interval $[\underline{a}, \overline{a}] = [\tilde{a} - \Delta_a, \tilde{a} + \Delta_a]$ describing the set of all desired values, and
- an interval $[\underline{b}, \overline{b}] = [\tilde{b} - \Delta_b, \tilde{b} + \Delta_b]$ describing the set of all the values which are possible for a given decision.

The Formula That We Need to Justify. Let us describe an explicit expression for the formula that we need to justify – the formula describing the degree d to which the proposed decision leads to the desired result.

The above procedure is symmetric with respect to changing a and b. So, if necessary, we can swap a and b. Thus, without losing generality, we can assume that $\tilde{a} \leq \tilde{b}$.

One can prove that the maximum d of the function $\min(\mu_a(x), \mu_i(x))$ is attained when the values $\mu_a(x)$ and $\mu_i(x)$ are equal, i.e., at a point x_m for which $\mu_a(x_m) = \mu_i(x_m)$. Indeed, if, e.g., $\mu_a(x_m) > \mu_i(x_m)$, then $\min(\mu_a(x_m), \mu_i(x_m)) = \mu_i(x_m)$. In this case, we cannot have $\mu_i(x_m) = 1$, so we must have $\mu_i(x_m) < 1$. In this case, by modifying x_m a little bit, we can increase the value $\mu_i(x)$ and thus, achieve a larger value of the $\min(\mu_a(x), \mu_i(x))$ – which contradicts to our assumption that the function attains maximum at x_m. Similarly, the maximum cannot be attained when $\mu_a(x_m) < \mu_i(x_m)$, so it has to be attained when $\mu_a(x_m) = \mu_i(x_m)$.

In this case, the desired maximum d is equal to $d = \mu_a(x_m) = \mu_i(x_m)$.

Since $\tilde{a} \leq \tilde{b}$, the equality $\mu_a(x_m) = \mu_i(x_m)$ is attained when:

- the value on the decreasing part of the a-membership function $\mu_a(x)$ (that linearly goes from 1 at $\tilde{a}$ to 0 at $\tilde{a} + \Delta_a$) coincides with
- the value on the increasing part of the b-membership function $\mu_b(x)$ (that linearly goes from 0 at $\tilde{b} - \Delta_b$ to 1 for $\tilde{b}$).

By applying the general formula

$$y = y_1 + \frac{y_2 - y_1}{x_2 - x_2} \cdot (x - x_1)$$

for a straight line that takes value y_1 at $x = x_1$ and value y_2 at $x = x_2$, we conclude that

$$\mu_a(x) = 1 - \frac{x - \tilde{a}}{\Delta_a} \text{ and } \mu_b(x) = \frac{x - (\tilde{b} - \Delta_b)}{\Delta_b}.$$

Thus, the condition $\mu_a(x_m) = \mu_i(x_m)$ takes the form

$$1 - \frac{x_m - \tilde{a}}{\Delta_a} = \frac{x_m - (\tilde{b} - \Delta_b)}{\Delta_b}.$$

By opening the parentheses, we get

$$1 - \frac{x_m - \tilde{a}}{\Delta_a} = \frac{x_m - \tilde{b} + \Delta_b}{\Delta_b} = 1 - \frac{\tilde{b} - x_m}{\Delta_b},$$

or, equivalently,

$$\frac{x_m - \tilde{a}}{\Delta_a} = \frac{\tilde{b} - x_m}{\Delta_b}.$$

Multiplying both side by Δ_a and Δ_b, we get

$$(x_m - \tilde{a}) \cdot \Delta_b = (\tilde{b} - x_m) \cdot \Delta_a.$$

Moving all the terms containing the unknown x_m into the left-hand side and all the other terms into the right-hand side, we get

$$x_m \cdot (\Delta_a + \Delta_b) = \tilde{a} \cdot \Delta_b + \tilde{b} \cdot \Delta_a,$$

hence

$$x_m = \frac{\tilde{a} \cdot \Delta_b + \tilde{b} \cdot \Delta_a}{\Delta_a + \Delta_b}.$$

Substituting this value into the formula for $\mu_a(x)$, we conclude that the desired maximum value d is equal to

$$d = \mu_a(x_m) = 1 - \frac{x_m - \tilde{a}}{\Delta_a}.$$

Here,

$$x_m - \tilde{a} = \frac{\tilde{a} \cdot \Delta_b + \tilde{b} \cdot \Delta_a}{\Delta_a + \Delta_b} - \tilde{a}.$$

By moving both terms in the right-hand side to the common denominator, we conclude that

$$x_m - \tilde{a} = \frac{\tilde{a} \cdot \Delta_b + \tilde{b} \cdot \Delta_a - \tilde{a} \cdot \Delta_a - \tilde{a} \cdot \Delta_b}{\Delta_a + \Delta_b} = \frac{\tilde{b} \cdot \Delta_a - \tilde{a} \cdot \Delta_a}{\Delta_a + \Delta_b}.$$

Thus,

$$d = 1 - \frac{\tilde{b} - \tilde{a}}{\Delta_a + \Delta_b}. \tag{3.1}$$

This is the formula that we need to justify.

Our Main Idea. If we knew the exact values of a and b, then we would be able to conclude that either $a = b$, or $a < b$, or $b < a$.

In reality, we know the values a and b with uncertainty. So, even if the actual values a and b are the same, we may get approximate values which are different; it is reasonable to assume that if the actual values are the same, then we have the same probability $\text{Prob}(a > b)$ and $\text{Prob}(b > a)$ of observing $a > b$ and $a < b$ are the same, i.e., that both these probabilities are equal to $1/2$. If the probabilities that $a > b$ and that $a < b$ differ, this is an indication that the actual value a and b are different,

Thus, it is reasonable to use the difference

$$|\text{Prob}(a > b) - \text{Prob}(b > a)|$$

as the degree to which a and b may be different.

How We Can Estimate the Probabilities $\text{Prob}(a > b)$ **and** $\text{Prob}(b > a)$**.** If we know the exact values of a and b, then we can check the inequality $a > b$ by computing the difference $a - b$ and comparing this difference to 0.

In real life, as we have mentioned, we only know a and b with interval uncertainty, i.e., we only know that

$$a \in [\tilde{a} - \Delta_a, \tilde{a} + \Delta_a] \text{ and } b \in [\tilde{b} - \Delta_b, \tilde{b} + \Delta_b].$$

In this case, we do not know the exact value of the difference $a - b$, we only know the range of possible values of this difference.

Such a range can be computed by using interval arithmetic; see, e.g., [49]. Namely:

- the smallest possible value of the difference $a - b$ is attained when a attains its smallest possible value $\tilde{a} - \Delta_a$ and b attains its largest possible value $\tilde{b} + \Delta_b$; the resulting difference is equal to

$$\tilde{a} - \Delta_a - (\tilde{b} + \Delta_b) = (\tilde{a} - \tilde{b}) - (\Delta_a + \Delta_b);$$

- the largest possible value of the difference $a - b$ is attained when a attains its largest possible value $\tilde{a} + \Delta_a$ and b attains its smallest possible value $\tilde{b} - \Delta_b$; the resulting difference is equal to

$$\tilde{a} + \Delta_a - (\tilde{b} - \Delta_b) = (\tilde{a} - \tilde{b}) + (\Delta_a + \Delta_b).$$

Thus, the only thing we know about the difference $a - b$ is that this difference belongs to the interval

$$[(\tilde{a} - \tilde{b}) - (\Delta_a + \Delta_b), (\tilde{a} - \tilde{b}) + (\Delta_a + \Delta_b)].$$

We do not have any reason to assume that some values from this interval are more probable and some are less probable. It is therefore reasonable to assume that all the values from this interval are equally probable, i.e., that the corresponding probability distribution is *uniform* on this interval.

Comment. This argument is widely used in data processing; it is called *Laplace principle of indifference* or *Laplace principle of insufficient reason*. Its most adequate mathematical description is the Maximum Entropy approach (see, e.g, [29]) – widely used in statistics – according to which, if several different probability distributions $\rho(x)$ are consistent with our knowledge, we should select the *least informative* one, i.e., the one for which the entropy $S = -\int \rho(x) \cdot \ln(\rho(x)) \, dx$ is the largest possible. In our case, all we know about the corresponding probability density function $\rho(x)$ is that it is located on a given interval $[\underline{c}, \overline{c}]$, i.e., that $\rho(x) = 0$ outside this interval. Thus, in all integrations containing $\rho(x)$, we can skip the parts where this function is 0 and only consider values from the interval $[\underline{c}, \overline{c}]$. So, to find the appropriate distribution, we must maximize the entropy

$$S = -\int_{\underline{c}}^{\overline{c}} \rho(x) \cdot \ln(\rho(x)) \, dx \rightarrow \max$$

under the constraints that $\rho(x) \geq 0$ and

$$\int_{\underline{c}}^{\overline{c}} \rho(x) \, dx = 1.$$

By using Lagrange multiplier method, we can reduce this constraint optimization to unconstrained optimization problem

$$-\int_{\underline{c}}^{\overline{c}} \rho(x) \cdot \ln(\rho(x)) \, dx + \lambda \cdot \left(\int_{\underline{c}}^{\overline{c}} \rho(x) \, dx - 1 \right) \rightarrow \max .$$

Differentiating the objective function with respect to $\rho(x)$, we conclude that

$$-\rho(x) - 1 + \lambda = 0,$$

i.e., that $\ln(\rho(x)) = \lambda - 1$ and thus, $\rho(x) = \exp(\lambda - 1)$. This value is the same for all $x \in [\underline{c}, \overline{c}]$, so we indeed get a uniform distribution.

How We Can Estimate the Probabilities Prob$(a > b)$ **and** Prob$(b > a)$ **(cont-d).** In our approach:

- as an estimate for the probability Prob$(a > b)$, we take the probability Prob$(a - b > 0)$ that the difference $a - b$ is positive, and
- as an estimate for the probability Prob$(a < b)$, we take the probability Prob$(a - b < 0)$ that the difference $a - b$ is negative.

Now that we have assumed that the probability distribution on the set of all the values $a - b$ is uniformly distributed on the interval

$$[(\widetilde{a} - \widetilde{b}) - (\Delta_a + \Delta_b), (\widetilde{a} - \widetilde{b}) + (\Delta_a + \Delta_b)],$$

we can find the numerical values for both probabilities. Namely,

- values $a - b > 0$ form a subinterval

$$(0, (\widetilde{a} - \widetilde{b}) + (\Delta_a + \Delta_b)];$$

- values $a - b < 0$ form a subinterval

$$[(\widetilde{a} - \widetilde{b}) - (\Delta_a + \Delta_b), 0).$$

In a uniform distribution, the probability to be in a subinterval is proportional to the width of this subinterval. The coefficient of proportionality can be found from the condition that the overall probability to be in the entire interval is equal to 1. Thus, when we have a uniform distribution on an arbitrary interval $[\underline{c}, \overline{c}]$, the probability p to be in a subinterval $[\underline{d}, \overline{d}] \subseteq [\underline{c}, \overline{c}]$ is equal to the ratio of the widths of these intervals:

$$p = \frac{\overline{d} - \underline{d}}{\overline{c} - \underline{c}}.$$

In our case, the width of the big interval is equal to

$$((\tilde{a} - \tilde{b}) + (\Delta_a + \Delta_b)) - ((\tilde{a} - \tilde{b}) - (\Delta_a + \Delta_b)) = 2 \cdot (\Delta_a + \Delta_b),$$

and thus, the probabilities $\text{Prob}(a > b)$ and $\text{Prob}(a < b)$ are equal to

$$\text{Prob}(a > b) = \frac{(\tilde{a} - \tilde{b}) + (\Delta_a + \Delta_b)}{2 \cdot (\Delta_a + \Delta_b)};$$

$$\text{Prob}(a < b) = \frac{(\Delta_a + \Delta_b) - (\tilde{a} - \tilde{b})}{2 \cdot (\Delta_a + \Delta_b)}.$$

So, the desired difference $\text{Prob}(a > b) - \text{Prob}(b > a)$ takes the form

$$\text{Prob}(a > b) - \text{Prob}(b > a) = \frac{(\tilde{a} - \tilde{b}) + (\Delta_a + \Delta_b)}{2 \cdot (\Delta_a + \Delta_b)} - \frac{(\Delta_a + \Delta_b) - (\tilde{a} - \tilde{b})}{2 \cdot (\Delta_a + \Delta_b)} =$$

$$\frac{2 \cdot (\tilde{a} - \tilde{b})}{2 \cdot (\Delta_a + \Delta_b)} = \frac{\tilde{a} - \tilde{b}}{\Delta_a + \Delta_b}.$$

Since $\tilde{a} \geq \tilde{b}$, we can conclude that

$$|\text{Prob}(a > b) - \text{Prob}(b > a)| = \frac{\tilde{a} - \tilde{b}}{\Delta_a + \Delta_b}. \tag{3.2}$$

Conclusion. By comparing:

- the above formula (3.1) for the degree d to which the alternative a fits the fuzzy target b
- with the formula (3.2) for the probability

$$|\text{Prob}(a > b) - \text{Prob}(b > a)|$$

with which the alternative a and the fuzzy target b are different,

we can see that

$$d + |\text{Prob}(a > b) - \text{Prob}(b > a)| = 1.$$

Thus, the degree d can be described, in reasonable probabilistic terms, as

$$d = 1 - |\text{Prob}(a > b) - \text{Prob}(b > a)|.$$

We have therefore produced a new justification for the above complex formula for d, the justification that does not use any simplifying assumptions and which is, therefore, applicable in the general case.

Chapter 4
Towards More Computationally Efficient Techniques for Processing Uncertainty

In this chapter, we consider challenges related to computational efficiency of uncertainty propagation.

One of the main reasons why the corresponding computations often take a lot of time is that we need to process a large amount of data. So, a natural way to speed up data processing is:

- to divide the large amount of data into smaller parts,
- process each smaller part separately, and then
- to combine the results of data processing.

In particular, when we are processing huge amounts of heterogenous data, it makes sense to first process different types of data type-by-type and then to fuse the resulting models. This idea is explored in the first sections of this Chapter.

Even with this idea in place, even when all reasonable algorithmic speed-up ideas have been implemented, the computation time is still often too long. In such situations, the only remaining way to speed up computations is to use different hardware speed-up ideas. Such ideas range from currently available (like parallelization) to more futuristic ideas like quantum computing. While parallelization has been largely well-researched, the use of future techniques (such as quantum computing) in data processing and uncertainty estimation is still largely an open problem. In the last section, we show how quantum computing can be used to speed up the corresponding computations.

The results from this chapter were first published in [40, 52, 53].

4.1 Model Fusion: A Way to Make Uncertainty Processing More Computationally Efficient

Need for Model Fusion. In many areas of science and engineering, we have different sources of data.

For example, as we have mentioned in Chapter 1, in geophysics, there are many sources of data for Earth models:

© Springer International Publishing Switzerland 2015
C. Servin & V. Kreinovich, *Propagation of Interval & Probabilistic Uncertainty*,
Studies in Systems, Decision and Control 15, DOI: 10.1007/978-3-319-12628-9_4

- first-arrival passive seismic data (from actual earthquakes);
- first-arrival active seismic data (from seismic experiments using man-made sources);
- gravity data; and
- surface waves.

Datasets coming from different sources provide complimentary information. For example, different geophysical datasets contain different information on earth structure. In general:

- some of the datasets provide better accuracy and/or spatial resolution in some spatial areas and in some depths, while
- other datasets provide a better accuracy and/or spatial resolution in other areas or depths.

For example:

- each measured gravity anomaly at a point is the result of the density distribution over a relatively large region of the earth, so estimates based on gravity measurements have (relatively) low spatial resolution;
- in contrast, each seismic data point (arrival time) comes from a narrow trajectory (ray) a seismic wave travels within the earth, so the spatial resolution corresponding to this data is much higher.

It is therefore desirable to combine data from different datasets.

The ideal approach would be to use all the datasets to produce a single model. At present, however, in many research areas – including geophysics – there are no efficient algorithms for simultaneously processing all the different datasets.

To solve this problem, we can use a natural solution: process different datasets separately, and then fuse all the *models* coming from different datasets.

Data Fusion as the Simplest Case of Model Fusion. Different models provide us with different $\widetilde{x}^{(1)}, \ldots, \widetilde{x}^{(n)}$ of the same quantity x. To combine these estimates into a single estimate, we can use data fusion techniques described in Chapter 1.

Need to Go Beyond Data Fusion. In many practical situations, estimates coming from different models have not only different accuracy, but also different spatial resolution. For example, in the geosciences,

- seismic data leads to estimates of the density at different locations and depths which have higher spatial resolution (based on an empirical relationship between density and seismic velocity), while
- gravity data leads to estimates of the same densities which have lower spatial resolution.

It is therefore necessary to go beyond data fusion, and to explicitly take different spatial resolution into account when fusing different models.

Towards Precise Formulation of the Problem. Estimates with higher spatial (spatio-temporal) resolution mean that we estimate the values corresponding to small spatial (spatio-temporal) cells. An estimate with a lower spatial resolution means that its results are affected by several neighboring spatial cells, i.e., that we are estimating, in effect, a weighted average of the values in several neighboring cells.

In precise terms:

- we have resolution estimates $\widetilde{x}_1, \ldots, \widetilde{x}_n$ of the values $x_1, \ldots, x_n$ within several small spatial cells; these estimates correspond to models with a higher spatial resolution
- we also have estimates $\widetilde{X}_j$ for the weighted averages $X_j = \sum_{i=1}^{n} w_{j,i} \cdot x_i$; these estimates correspond to models with a lower spatial resolution.

Comment. In this chapter, we assume that we know the values of the weights $w_{j,i}$. This assumption makes perfect sense for geophysical problems, because in these problems, these weights are indeed known. For example:

- We know how exactly the gravity at a given point depends on the densities at different spatial locations.
- We know how the travel time depends on the density distribution: specifically, we know how exactly the travel time of a seismic signal depends on the velocity distribution, and we know an empirical velocity-density relationship.

In some applications, however, the corresponding weights are only approximately known. In such situations, when fusing the models, we must also take into account the uncertainty with which we know these weights. For these applications, it is desirable to extend our techniques – to accommodate such more complex situations.

What We Do in This Chapter. We are interested in the values x_i. So, based on the estimates $\widetilde{x}_i$ and $\widetilde{x}$, we must provide more accurate estimates for x_i.

For example, in the geophysical example, we are interested in the values of the densities x_i.

In this chapter, we describe how to fuse estimates with different accuracy and spatial resolution:

- In the case of probabilistic uncertainty, we use the Least Squares Method to derive explicit formulas for combining the estimates $\widetilde{x}_i$ and $\widetilde{X}_j$.
- In the case of interval uncertainty, we provide an efficient algorithm for estimating the ranges of x_i.

4.2   Model Fusion: Case of Probabilistic Uncertainty

4.2.1   General Case

Main Idea. Our solution to the model fusion problem is to take into account several different types of approximate equalities:

- Each estimate $\widetilde{x}_i$ from a model with a high spatial resolution is approximately equal to the actual value x_i in the corresponding (smaller size) cell i, with the known accuracy $\sigma_{h,i}$:

$$\widetilde{x}_i \approx x_i.$$

- Each estimate $\widetilde{X}_j$ from (one of the) models with a lower spatial resolution is approximately equal to the weighted average of values of all the smaller cells $x_{i(1,j)}, \ldots, x_{i(k_j,j)}$ within the corresponding larger size cell, with a known accuracy $\sigma_{l,j}$:

$$\widetilde{X}_j \approx \sum_i w_{j,i} \cdot x_i,$$

for known weights $w_{j,i} \geq 0$ for which $\sum_{i=1}^{n} w_{j,i} = 1$. In the simple case when these weights are equal, we get

$$\widetilde{X}_j \approx \frac{x_{i(1,j)} + \ldots + x_{i(k_j,j)}}{k_j}.$$

- We usually have a prior knowledge of the values x_i. It is reasonable to assume that this knowledge can also be described by a normal distribution, with the mean $x_{pr,i}$ and the standard deviation $\sigma_{pr,i}$:

$$x_i \approx x_{pr,i}.$$

(The case when for some i, we have no prior information at all is equivalent to setting $\sigma_{pr,i} = \infty$.)

- Finally, each estimate $\widetilde{X}_j$ from a model with a lower spatial resolution is approximately equal to the value within each of the constituent smaller size cells $x_{i(l,j)}$, with the accuracy corresponding to the (empirical) standard deviation $\sigma_{e,j}$ of the smaller-cell values within the larger cell:

$$\widetilde{X}_j \approx x_{i(l,j)},$$

where $\sigma_{e,j}^2 \stackrel{\text{def}}{=} \dfrac{1}{k_j} \cdot \sum_{l=1}^{k_j} \left(\widetilde{x}_{i(l,j)} - E_j \right)^2$, and $E_j \stackrel{\text{def}}{=} \dfrac{1}{k_j} \cdot \sum_{l=1}^{k_j} \widetilde{x}_{i(l,j)}$.

We then use the Least Squares technique to combine these approximate equalities, and find the desired combined values x_i by minimizing the resulting sum of weighted squared differences.

Relation between Different Standard Deviations. As we have mentioned earlier, there is usually a trade-off between accuracy and spatial resolution:

- if we want to estimate the value of the desired quantity with a higher spatial resolution, i.e., the value corresponding to a small spatial location, then we get lower accuracy, i.e., higher values of the standard deviation $\sigma_{h,i}$;
- on the other hand, if we are satisfied with a lower spatial resolution, i.e., with the fact that the estimated value corresponds to a larger spatial area, then we can get higher accuracy, i.e., lower values of the standard deviation $\sigma_{l,j} \ll \sigma_{h,i}$.

From the mathematical viewpoint, this trade-off makes sense. In principle, as an estimate for a model with a low spatial resolution, we can take the average of the values corresponding to high spatial resolution, and averaging usually decreases the approximation error: $\sigma_{l,j} \ll \sigma_{h,i} \ll \sigma_{e,j}$.

Comment. It should be mentioned that while usually, higher spatial resolution estimates have lower accuracy, sometimes, a higher-resolution model has more accuracy in some places. For example, in the geosciences,

- the measurements from a borehole provide the most accurate estimates of the corresponding quantities,
- and for these measurements, the spatial location is also known with a very high accuracy.

Resulting Formulas: General Case. According to the Least Squares approach, in the general case, we minimize the following expression:

$$\sum_{i=1}^{n} \frac{(x_i - \widetilde{x}_i)^2}{\sigma_{h,i}^2} + \sum_{j=1}^{m} \frac{1}{\sigma_{l,j}^2} \cdot \left(\widetilde{X}_j - \sum_{i=1}^{n} w_{j,i} \cdot x_i \right)^2 +$$

$$\sum_{i=1}^{n} \frac{(x_i - x_{pr,i})^2}{\sigma_{pr,i}^2} + \sum_{j=1}^{m} \sum_{l=1}^{k_j} \frac{(\widetilde{X}_j - x_{i(l,j)})^2}{\sigma_{e,j}^2}.$$

In this general case, differentiation with respect to x_i leads to the following system of linear equations:

$$\frac{x_i - \widetilde{x}_i}{\sigma_{h,i}^2} + \sum_{j:j \ni i} \frac{w_{j,i}}{\sigma_{l,j}^2} \cdot \left(\sum_{i'=1}^{n} w_{j,i'} \cdot x_{i'} - \widetilde{X}_j \right) + \frac{x_i - x_{pr,i}}{\sigma_{pr,i}^2} + \sum_{j:j \ni i} \frac{x_i - \widetilde{X}_j}{\sigma_{e,j}^2} = 0,$$

where $j \ni i$ means that the j-th estimate corresponding to a model with a low spatial resolution covers the i-th cell.

Towards Simplification: Fusing Prior Estimates with Estimates from a Model with a High Spatial Resolution. For each cell i for which we have both a prior estimate $x_{pr,i}$ and an estimate $\widetilde{x}_i$ from a model with a higher spatial resolution, we can fuse these two estimates by using the above-described standard data fusion technique. As a result, instead of the two terms

$$\sigma_{h,i}^{-2} \cdot (x_i - \tilde{x}_i) + \sigma_{pr,i}^{-2} \cdot (x_i - x_{pr,i}),$$

we have a single term $\sigma_{f,i}^{-2} \cdot (x_i - x_{f,i})$, where $x_{f,i} \overset{\text{def}}{=} \dfrac{\tilde{x}_i \cdot \sigma_{h,i}^{-2} + x_{pr,i} \cdot \sigma_{pr,i}^{-2}}{\sigma_{h,i}^{-2} + \sigma_{pr,i}^{-2}}$ and $\sigma_{f,i}^{-2} \overset{\text{def}}{=}$

$\sigma_{h,i}^{-2} + \sigma_{pr,i}^{-2}$. We can use the same formula if we only have a high spatial resolution estimate or if we only have a prior estimate:

- If we only have a high spatial resolution estimate but no prior estimate, then we should take $\sigma_{pr,i}^{-2} = 0$ (i.e., $\sigma_{pr,i} = \infty$).
- If we only have a prior estimate but no high spatial resolution estimate, then we should take $\sigma_{h,i}^{-2} = 0$ (i.e., $\sigma_{h,i} = \infty$).

As a result of this fusion, we get the following simplified formulas.

Resulting Formulas: Simplified Equations

$$\frac{x_i - x_{f,i}}{\sigma_{f,i}^2} + \sum_{j:j \ni i} \frac{w_{j,i}}{\sigma_{l,j}^2} \cdot \left(\sum_{i'=1}^{n} w_{j,i'} \cdot x_{i'} - \tilde{X}_j \right) + \sum_{j:j \ni i} \frac{x_i - \tilde{X}_j}{\sigma_{e,j}^2} = 0.$$

How to Solve This System of Linear Equations. We can use known algorithms for solving this system of linear equations.

It is worth mentioning that usually, these algorithms require that we represent the system in the standard form $Ax = b$. To represent our system of equations in this form, we need to move all the terms that do not contain unknowns to the right-hand side.

4.2.2 Simplest Case

Description. Let us now consider the simplest case, when when we have exactly one estimate $\tilde{X}_1$ from a model with a low spatial resolution. In general, we only have prior estimates and the estimates with high spatial resolution for *some* of the cells.

This situation is typical in geosciences: e.g.,

- we have an estimate originated from the gravity measurements (with a lower spatial resolution) which covers a huge area in depth, and
- we have estimates originated from seismic measurements (corresponding to higher spatial resolution) which only cover depths above the Moho surface (the base of the earth's crust).

For convenience, let us number the cells in such a way that the cells for which we have either prior estimates or estimates from a high spatial resolution model come first. Let h denote the total number of such cells.

This means that as the result of combining prior estimates and estimates corresponding to high spatial resolution model(s), we have h values $x_{f,1}, x_{f,2}, \ldots, x_{f,h}$.

Derivation. In this case, the above system of linear equations takes the following form: for $i = 1, \ldots, h$, we have

$$\sigma_{f,i}^{-2} \cdot (x_i - x_{f,i}) + \frac{1}{\sigma_{l,1}^2} \cdot w_{1,i} \cdot \left(\sum_{i'} w_{1,i'} \cdot x_{i'} - \widetilde{X}_1 \right) + \frac{1}{\sigma_{e,1}^2} (x_i - \widetilde{X}_1) = 0;$$

and for $i > h$, we have

$$\frac{1}{\sigma_{l,1}^2} \cdot w_{1,i} \cdot \left(\sum_{i'} w_{1,i'} \cdot x_{i'} - \widetilde{X}_1 \right) + \frac{1}{\sigma_{e,1}^2} (x_i - \widetilde{X}_1) = 0.$$

For $i \leq h$, multiplying both sides by $\sigma_{f,i}^2$, we conclude that

$$x_i - x_{f,i} + \frac{\sigma_{f,i}^2}{\sigma_{l,1}^2} \cdot w_{1,i} \cdot \left(\sum_{i'} w_{1,i'} \cdot x_{i'} - \widetilde{X}_1 \right) + \frac{\sigma_{f,i}^2}{\sigma_{e,1}^2} \cdot (x_i - \widetilde{X}_1) = 0.$$

If we introduce an auxiliary variable $\mu \overset{\text{def}}{=} \dfrac{1}{\sigma_{l,1}^2} \cdot \left(\sum_{i'} w_{1,i'} \cdot x_{i'} - \widetilde{X}_1 \right)$, we get the equation

$$x_i - x_{f,i} + w_{1,i} \cdot \sigma_{f,i}^2 \cdot \mu + \frac{\sigma_{f,i}^2}{\sigma_{e,1}^2} \cdot (x_i - \widetilde{X}_1) = 0.$$

By keeping terms proportional to x_i in the left-hand side and by moving all the other terms to the right-hand side, we get $\left(1 + \dfrac{\sigma_{f,i}^2}{\sigma_{e,1}^2} \right) \cdot x_i = x_{f,i} - w_{1,i} \cdot \sigma_{f,i}^2 \cdot \mu + \dfrac{\sigma_{f,i}^2}{\sigma_{e,1}^2} \cdot \widetilde{X}_1,$ hence

$$x_i = \frac{x_{f,i}}{1 + \dfrac{\sigma_{f,i}^2}{\sigma_{e,1}^2}} - \frac{w_{1,i} \cdot \sigma_{f,i}^2}{1 + \dfrac{\sigma_{f,i}^2}{\sigma_{e,1}^2}} \cdot \mu + \widetilde{X}_1 \cdot \frac{\dfrac{\sigma_{f,i}^2}{\sigma_{e,1}^2}}{1 + \dfrac{\sigma_{f,i}^2}{\sigma_{e,1}^2}}.$$

For $i > h$, we similarly get $x_i - \widetilde{X}_1 + w_{1,i} \cdot \sigma_{e,1}^2 \cdot \mu = 0$, hence $x_i = \widetilde{X}_1 - w_{1,i} \cdot \sigma_{e,1}^2 \cdot \mu$.

To make this expression practically useful, we must describe μ in terms of the given values $\widetilde{x}_i$ and $\widetilde{X}_1$. Since μ is defined in terms of the weighted average of the values x_i, let us compute the weighted average of the above expressions for x_i: $\sum\limits_{i=1}^{n} w_{1,i} \cdot x_i = \sum\limits_{i=1}^{h} w_{1,i} \cdot x_i + \sum\limits_{i=h+1}^{n} w_{1,i} \cdot x_i$, where

$$\sum_{i=1}^{h} w_{1,i} \cdot x_i = \sum_{i=1}^{h} \frac{w_{1,i} \cdot x_{f,i}}{1 + \dfrac{\sigma_{f,i}^2}{\sigma_{e,1}^2}} - \mu \cdot \sum_{i=1}^{h} \frac{w_{1,i}^2 \cdot \sigma_{f,i}^2}{1 + \dfrac{\sigma_{f,i}^2}{\sigma_{e,1}^2}} + \widetilde{X}_1 \cdot \sum_{i=1}^{h} \frac{w_{1,i} \cdot \dfrac{\sigma_{f,i}^2}{\sigma_{e,1}^2}}{1 + \dfrac{\sigma_{f,i}^2}{\sigma_{e,1}^2}}.$$

Similarly,

$$\sum_{i=h+1}^{n} w_{1,i} \cdot x_i = \left(\sum_{i=h+1}^{n} w_{1,i} \right) \cdot \tilde{X}_1 - \left(\sum_{i=h+1}^{n} w_{1,i}^2 \right) \cdot \frac{\sigma_{e,1}^2}{\sigma_{l,1}^2} \cdot \mu.$$

By adding these two sums and subtracting $\tilde{X}_1$, we conclude that

$$\sigma_{l,1}^2 \cdot \mu = \sum_{i=1}^{n} w_{1,i} \cdot x_i - \tilde{X}_1 = \sum_{i=1}^{h} w_{1,i} \cdot x_i + \sum_{i=h+1}^{n} w_{1,i} \cdot x_i - \tilde{X}_1 =$$

$$\sum_{i=1}^{h} \frac{w_{1,i} \cdot x_{f,i}}{1 + \frac{\sigma_{f,i}^2}{\sigma_{e,1}^2}} - \mu \cdot \sum_{i=1}^{h} \frac{w_{1,i}^2 \cdot \sigma_{f,i}^2}{1 + \frac{\sigma_{f,i}^2}{\sigma_{e,1}^2}} + \tilde{X}_1 \cdot \sum_{i=1}^{h} \frac{w_{1,i} \cdot \frac{\sigma_{f,i}^2}{\sigma_{e,1}^2}}{1 + \frac{\sigma_{f,i}^2}{\sigma_{e,1}^2}} +$$

$$\left(\sum_{i=h+1}^{n} w_{1,i} \right) \cdot \tilde{X}_1 - \left(\sum_{i=h+1}^{n} w_{1,i}^2 \right) \cdot \sigma_{e,1}^2 \cdot \mu - \tilde{X}_1.$$

Since $\sum_{i=1}^{n} w_{1,i} = \sum_{i=1}^{h} w_{1,i} + \sum_{i=h+1}^{n} w_{1,i} = 1$, we conclude that

$$\left(\sum_{i=h+1}^{n} w_{1,i} \right) \cdot \tilde{X}_1 - \tilde{X}_1 = - \left(\sum_{i=1}^{h} w_{1,i} \right) \cdot \tilde{X}_1$$

thus,

$$\tilde{X}_1 \cdot \sum_{i=1}^{h} \frac{w_{1,i} \cdot \frac{\sigma_{f,i}^2}{\sigma_{e,1}^2}}{1 + \frac{\sigma_{f,i}^2}{\sigma_{e,1}^2}} + \left(\sum_{i=h+1}^{n} w_{1,i} \right) \cdot \tilde{X}_1 - \tilde{X}_1 =$$

$$\tilde{X}_1 \cdot \sum_{i=1}^{h} \frac{w_{1,i} \cdot \frac{\sigma_{f,i}^2}{\sigma_{e,1}^2}}{1 + \frac{\sigma_{f,i}^2}{\sigma_{e,1}^2}} - \left(\sum_{i=1}^{h} w_{1,i} \right) \cdot \tilde{X}_1 = -\tilde{X}_1 \cdot \sum_{i=1}^{h} \frac{w_{1,i}}{1 + \frac{\sigma_{f,i}^2}{\sigma_{e,1}^2}}.$$

So, the equation for μ takes the following simplified form:

$$\sigma_{l,1}^2 \cdot \mu =$$

$$\sum_{i=1}^{h} \frac{w_{1,i} \cdot x_{f,i}}{1 + \frac{\sigma_{f,i}^2}{\sigma_{e,1}^2}} - \mu \cdot \sum_{i=1}^{h} \frac{w_{1,i}^2 \cdot \sigma_{f,i}^2}{1 + \frac{\sigma_{f,i}^2}{\sigma_{e,1}^2}} - \tilde{X}_1 \cdot \sum_{i=1}^{h} \frac{w_{1,i}}{1 + \frac{\sigma_{f,i}^2}{\sigma_{e,1}^2}} - \left(\sum_{i=h+1}^{n} w_{1,i}^2 \right) \cdot \sigma_{e,1}^2 \cdot \mu.$$

By moving all terms containing μ to the left-hand side and all other terms to the right-hand side, we get an explicit equation for μ. So, we arrive at the following formulas.

Resulting Formulas. First, we compute the auxiliary value μ as $\mu = \dfrac{N}{D}$, where

$$N = \sum_{i=1}^{h} \frac{w_{1,i} \cdot (x_{f,i} - \widetilde{X}_1)}{1 + \dfrac{\sigma_{f,i}^2}{\sigma_{e,1}^2}} \quad \text{and} \quad D = \sigma_{l,1}^2 + \sum_{i=1}^{h} \frac{w_{1,i}^2 \cdot \sigma_{f,i}^2}{1 + \dfrac{\sigma_{f,i}^2}{\sigma_{e,1}^2}} + \left(\sum_{i=h+1}^{n} w_{1,i}^2 \right) \cdot \sigma_{e,1}^2.$$

Then, we compute the desired estimates for x_i, $i = 1, \ldots, h$, as

$$x_i = \frac{x_{f,i}}{1 + \dfrac{\sigma_{f,i}^2}{\sigma_{e,1}^2}} - \frac{w_{1,i} \cdot \sigma_{f,i}^2}{1 + \dfrac{\sigma_{f,i}^2}{\sigma_{e,1}^2}} \cdot \mu + \widetilde{X}_1 \cdot \frac{\dfrac{\sigma_{f,i}^2}{\sigma_{e,1}^2}}{1 + \dfrac{\sigma_{f,i}^2}{\sigma_{e,1}^2}},$$

and the estimates x_i for $i = h+1, \ldots, n$ as $x_i = \widetilde{X}_1 - w_{1,i} \cdot \sigma_{e,1}^2 \cdot \mu$.

4.2.3 Numerical Example

Simplified Case: Description. To illustrate the above formulas, let us consider the simplest possible case, when we have exactly one estimate $\widetilde{X}_1$ from a lower spatial resolution model, and when:

- this estimate covers all n cells;
- all the weights are all equal $w_{1,i} = 1/n$;
- for each of n cells, there is an estimate corresponding to this cell that comes from a high spatial resolution model (i.e., $h = n$);
- all estimates coming from a high spatial resolution model have the same accuracy $\sigma_{h,i} = \sigma_h$;
- the estimate corresponding to a low spatial resolution model is much more accurate than the estimates corresponding to higher spatial resolution models $\sigma_{l,1} \ll \sigma_h$, so we can safely assume that $\sigma_l = 0$; and
- there is no prior information, so $\sigma_{pr,i} = \infty$ and thus, $x_{f,i} = \widetilde{x}_i$ and $\sigma_{f,i} = \sigma_h$.

To cover the cells for which there are no estimates from a high spatial resolution model, we added a heuristic rule that the estimate from a lower spatial resolution model is approximately equal to the value within each of the constituent smaller size cells, with the accuracy corresponding to the (empirical) standard deviation $\sigma_{e,j}$. In our simplified example, we have individual estimates for each cell, so there is no need for this heuristic rule. The corresponding heuristic terms in the general least squares approach are proportional to $\dfrac{1}{\sigma_{e,1}^2}$, so ignoring these terms is equivalent to taking $\sigma_{e,1}^2 = \infty$. Thus, we have $\dfrac{\sigma_{f,i}^2}{\sigma_{e,1}^2} = 0$ and $1 + \dfrac{\sigma_{f,i}^2}{\sigma_{e,1}^2} = 1$.

Because of this and because of the fact that $w_{1,i} = \dfrac{1}{n}$ and $x_{f,i} = \tilde{x}_i$, the formula for N takes the form

$$N = \sum_{i=1}^{n} \frac{1}{n} \cdot (\tilde{x}_i - \tilde{X}_1).$$

Opening parentheses and taking into account that the sum of n terms equal to $\dfrac{1}{n} \cdot \tilde{X}_1$ is simply $\tilde{X}_1$, we get

$$N = \frac{1}{n} \cdot \sum_{i=1}^{n} \tilde{x}_i - \tilde{X}_1.$$

Similarly, due to our simplifying assumptions $\sigma_{l,1} = 0$, $w_{1,i} = \dfrac{1}{n}$, $\sigma_{f,i} = \sigma_h$, $\sigma_{e,1} = 0$, and $h = n$, we have

$$D = \sum_{i=1}^{n} \left(\frac{1}{n}\right)^2 \cdot \sigma_h^2 = \frac{1}{n} \cdot \sigma_h^2.$$

Thus,

$$\mu = \frac{N}{D} = \frac{\dfrac{1}{n} \cdot \displaystyle\sum_{i=1}^{n} \tilde{x}_i - \tilde{X}_1}{\dfrac{1}{n} \cdot \sigma_h^2}.$$

The formula for x_i now turns into

$$x_i = \tilde{x}_i - \frac{1}{n} \cdot \sigma_h^2 \cdot \mu.$$

Substituting the above expression for μ, we conclude that

$$x_i = \tilde{x}_i - \lambda,$$

where

$$\lambda \stackrel{\text{def}}{=} \frac{1}{n} \cdot \sum_{i=1}^{n} \tilde{x}_i - \tilde{X}_1.$$

Numerical Example: Simplified Case. Let us assume that we have $n = 4$ cells, and that the high spatial resolution estimates for these cells are $\tilde{x}_1 = 2.0$, $\tilde{x}_2 = 3.0$, $\tilde{x}_3 = 5.0$ and $\tilde{x}_4 = 6.0$. We also assume that each of these estimates has the same accuracy $\sigma_h = 0.5$. Let us also assume that we have an estimate $\tilde{X}_1 = 3.7$ for the average X_1 of these four values. We assume that this estimate has a much higher accuracy $\sigma_l \ll \sigma_h$ so that we can, in effect, take $\sigma_l \approx 0$.

$\widetilde{x}_1 = 2.0$	$\widetilde{x}_2 = 3.0$
$\widetilde{x}_3 = 5.0$	$\widetilde{x}_4 = 6.0$

$\widetilde{X}_1 = 3.7$

Fig. 4.1 Higher and lower spatial resolution estimates

Since we assume that the low spatial resolution estimates are accurate ($\sigma_l \approx 0$), we therefore assume that the estimated quantity, i.e., the arithmetic average of the four cell values, is practically exactly equal to this estimate $\widetilde{X}_1 = 3.7$:

$$\frac{x_1 + x_2 + x_3 + x_4}{4} \approx 3.7.$$

For the high spatial resolution estimates $\widetilde{x}_i$, the average is slightly different:

$$\frac{\widetilde{x}_1 + \widetilde{x}_2 + \widetilde{x}_3 + \widetilde{x}_4}{4} = \frac{2.0 + 3.0 + 5.0 + 6.0}{4} = 4.0 \neq 3.7.$$

This difference is caused by the fact that, in contrast to accurate low spatial resolution estimates, higher spatial resolution measurements are much less accurate: the corresponding estimation error has a standard deviation $\sigma_h = 0.5$. We can therefore, as we described above, use the information from the low spatial resolution estimates to "correct" the high spatial resolution estimates.

In this particular example, since $\sigma_l \approx 0$, the correcting term takes the form

$$\lambda = \frac{\widetilde{x}_1 + \ldots + \widetilde{x}_n}{n} - \widetilde{X}_1 =$$

$$\frac{2.0 + 3.0 + 5.0 + 6.0}{4} - 3.7 = 4.0 - 3.7 = 0.3,$$

so the corrected ("fused") values x_i take the form:

$$x_1 = \widetilde{x}_1 - \lambda = 2.0 - 0.3 = 1.7; \quad x_2 = \widetilde{x}_2 - \lambda = 3.0 - 0.3 = 2.7;$$

$$x_3 = \widetilde{x}_3 - \lambda = 5.0 - 0.3 = 4.7; \quad x_4 = \widetilde{x}_4 - \lambda = 6.0 - 0.3 = 5.7;$$

For these corrected values, the arithmetic average is equal to

$$\frac{x_1 + x_2 + x_3 + x_4}{4} = \frac{1.7 + 2.7 + 4.7 + 5.7}{4} = 3.7,$$

i.e., exactly to the low spatial resolution estimate.

$\tilde{x}_1 = 1.7$	$\tilde{x}_2 = 2.7$
$\tilde{x}_3 = 4.7$	$\tilde{x}_4 = 5.7$

Fig. 4.2 The result of model fusion: simplified setting

Taking $\sigma_{e,j}$ into Account. What if, in the above numerical example, we take into account the requirement that the actual values in each cell are approximately equal to $\tilde{X}_1$, with the accuracy $\sigma_{e,1}$ equal to the empirical standard deviation?

In this case, the above formulas take the form

$$N = \frac{1}{1 + \dfrac{\sigma_h^2}{\sigma_{e,1}^2}} \cdot \left(\frac{\tilde{x}_1 + \ldots + \tilde{x}_n}{n} - \tilde{X}_1 \right)$$

and

$$D = \frac{1}{1 + \dfrac{\sigma_h^2}{\sigma_{e,1}^2}} \cdot \frac{1}{n} \cdot \sigma_h^2,$$

so we get the exact same expression for μ:

$$\mu = \frac{N}{D} = \frac{\dfrac{1}{n} \cdot \sum_{i=1}^{n} \tilde{x}_i - \tilde{X}_1}{\dfrac{1}{n} \cdot \sigma_h^2}.$$

The formulas for the fused values x_i are now somewhat more complex:

$$x_i = \frac{\tilde{x}_i - \lambda}{1 + \dfrac{\sigma_h^2}{\sigma_{e,1}^2}} + \tilde{X}_1 \cdot \frac{\dfrac{\sigma_h^2}{\sigma_{e,1}^2}}{1 + \dfrac{\sigma_h^2}{\sigma_{e,1}^2}}.$$

Taking $\sigma_{e,j}$ into Account: Numerical Example. We want to take into account the requirement that the actual values in each cell are approximately equal to $\tilde{X}_1$, with the accuracy $\sigma_{e,j}$ equal to the empirical standard deviation. In our example, the lower spatial resolution estimate $\tilde{X}_1$ covers all four cells. In this example, the above condition takes the form $x_i \approx \tilde{X}_1$, with the accuracy

$$\sigma_{e,1}^2 = \frac{1}{4} \cdot \sum_{i=1}^{4} (\tilde{x}_i - E_1)^2,$$

where

$$E_1 = \frac{1}{4} \cdot \sum_{i=1}^{4} \tilde{x}_i.$$

For our numerical example, as we have seen,

$$E_1 = \frac{1}{4} \cdot \sum_{i=1}^{4} \tilde{x}_i = \frac{\tilde{x}_1 + \tilde{x}_2 + \tilde{x}_3 + \tilde{x}_4}{4} = 4.0$$

and thus,

$$\sigma_{e,1}^2 = \frac{(2.0 - 4.0)^2 + (3.0 - 4.0)^2 + (5.0 - 4.0)^2 + (6.0 - 4.0)^2}{4} =$$

$$\frac{4 + 1 + 1 + 4}{4} = \frac{10}{4} = 2.5,$$

hence $\sigma_{e,1} \approx 1.58$.

Now, we can use the formula

$$x_i = \frac{1}{1 + \dfrac{\sigma_h^2}{\sigma_{e,1}^2}} \cdot (\tilde{x}_i - \lambda) + \frac{\dfrac{\sigma_h^2}{\sigma_{e,1}^2}}{1 + \dfrac{\sigma_h^2}{\sigma_{e,1}^2}} \cdot \tilde{X}_1$$

to find the corrected ("fused") values x_i. Here, $\sigma_h = 0.5$, $\sigma_{e,1}^2 = 2.5$, so

$$\frac{\sigma_h^2}{\sigma_{e,1}^2} = \frac{0.25}{2.5} = 0.1$$

and therefore, with two digit accuracy,

$$\frac{1}{1 + \dfrac{\sigma_h^2}{\sigma_{e,1}^2}} = \frac{1}{1.1} \approx 0.91$$

and

$$\frac{\dfrac{\sigma_h^2}{\sigma_{e,1}^2}}{1 + \dfrac{\sigma_h^2}{\sigma_{e,1}^2}} \cdot \tilde{X}_1 = \frac{0.1}{1.1} \cdot 3.7 \approx 0.34.$$

Therefore, we get

$$x_1 \approx 0.91 \cdot (2.0 - 0.3) + 0.34 \approx 1.89;$$

$$x_2 \approx 0.91 \cdot (3.0 - 0.3) + 0.34 \approx 2.79;$$

$$x_3 \approx 0.91 \cdot (5.0 - 0.3) + 0.34 \approx 4.62;$$

$$x_4 \approx 0.91 \cdot (6.0 - 0.3) + 0.34 \approx 5.53.$$

$\widetilde{x}_1 \approx 1.89$	$\widetilde{x}_2 \approx 2.79$
$\widetilde{x}_3 \approx 4.62$	$\widetilde{x}_4 \approx 5.53$

Fig. 4.3 The result of model fusion: general setting

The arithmetic average of these four values is equal to

$$\frac{x_1 + x_2 + x_3 + x_4}{4} \approx \frac{1.89 + 2.79 + 4.62 + 5.53}{4} \approx 3.71,$$

i.e., within our computation accuracy (since we performed all the computations with two digits after the decimal point) coincides with the lower spatial resolution estimate $\widetilde{X}_1 = 3.7$.

4.3 Model Fusion: Case of Interval Uncertainty

Main Idea. Our solution to the model fusion problem is to take into account three different types of approximate equalities:

- Each higher spatial resolution estimate $\widetilde{x}_i$ is approximately equal to the actual value x_i in the corresponding (smaller size) cell i, with the approximation error $x_i - \widetilde{x}_i$ bounded by the known value $\Delta_{h,i}$:

$$\widetilde{x}_i - \Delta_{h,i} \leq x_i \leq \widetilde{x}_i + \Delta_{h,i}.$$

- Each lower spatial resolution estimate $\widetilde{X}_j$ is approximately equal to the average of values of all the smaller cells $x_{i(1,j)}, \ldots, x_{i(k_j,j)}$ within the corresponding larger size cell, with the estimation error bounded by the known value $\Delta_{l,j}$:

$$\widetilde{X}_j - \Delta_{l,j} \leq \sum_i w_{j,i} \cdot x_i \leq \widetilde{X}_j + \Delta_{l,j}.$$

- Finally, we have prior bounds $\underline{x}_{pr,i}$ and $\overline{x}_{pr,i}$ on the values x_i, i.e., bounds for which

$$\underline{x}_{pr,i} \leq x_i \leq \overline{x}_{pr,i}.$$

Our objective is to find, for each $k = 1,\dots,n$, the range $[\underline{x}_k, \overline{x}_k]$ of possible values of x_k.

The estimates lead to a system of linear inequalities for the unknown values $x_1,\dots,x_n$. Thus, for each k, finding the corresponding endpoints $\underline{x}_k$ and $\overline{x}_k$ means optimizing the values x_k under linear constraints. This is a particular case of a general linear programming problem; see, e.g., [9]. So, we can use Linear Programming to find these bounds:

- the lower bound $\underline{x}_k$ can be obtained if we minimize x_k under the constraints

$$\widetilde{x}_i - \Delta_h \leq x_i \leq \widetilde{x}_i + \Delta_h, \quad i = 1,\dots,n;$$

$$\widetilde{X}_j - \Delta_l \leq \sum_i w_{j,i} \cdot x_i \leq \widetilde{X}_j + \Delta_l; \quad \underline{x}_{pr,i} \leq x_i \leq \overline{x}_{pr,i}.$$

- the upper bound $\overline{x}_k$ can be obtained if we maximize x_k under the same constraints.

Mathematical comment. For each i, the two constraints $\widetilde{x}_i - \Delta_h \leq x_i \leq \widetilde{x}_i + \Delta_h$ and $\underline{x}_{pr,i} \leq x_i \leq \overline{x}_{pr,i}$ can be combined into a single set of constraints:

$$x_i^- \leq x_i \leq x_i^+,$$

where

$$x_i^- \overset{\text{def}}{=} \max(\widetilde{x}_i - \Delta_h, \underline{x}_{pr,i}); \quad x_i^+ \overset{\text{def}}{=} \min(\widetilde{x}_i + \Delta_h, \overline{x}_{pr,i}).$$

Simplest Case: Description. Let us consider the simplest case when we have a single lower spatial resolution estimate $\widetilde{X}_1$. In this case, the linear constraints take the form $x_i^- \leq x_i \leq x_i^+$ and

$$\widetilde{X}_1 - \Delta_l \leq \sum_{i=1}^{n} w_{1,i} \cdot x_i \leq \widetilde{X}_1 + \Delta_l.$$

Comment. This general expression also includes the case when some cells are not covered by the estimate $\widetilde{X}_1$: for the values corresponding to these cells, we simply have $w_{1,i} = 0$.

Simplest Case: Derivation. Let us select a variable x_k, $k = 1,\dots,n$, and let us check which values of x_k are possible.

If the k-th cell is not affected by the estimate $\widetilde{X}_1$, i.e., if $w_{1,k} = 0$, then the only restrictions on x_k come from the prior bounds on x_k and from the higher spatial resolution estimates. Thus, for such a cell, the set of possible values is the interval $[x_k^-, x_k^+]$.

Let us now consider the case when the k-th cell is affected by the estimate $\widetilde{X}_1$, i.e., when $w_{1,k} > 0$. In this case, a possible value x_k must be within the interval $[x_k^-, x_k^+]$, and for the remaining variables x_i, $i = 1, \ldots, k-1, k+1, \ldots, n$, the resulting system of inequalities $x_i^- \leq x_i \leq x_i^+$ and

$$\widetilde{X}_1 - \Delta_l - w_{1,k} \cdot x_k \leq \sum_{i \neq k} w_{1,i} \cdot x_i \leq \widetilde{X}_1 + \Delta_l - w_{1,k} \cdot x_k$$

must be consistent.

All the weights $w_{1,i}$ are non-negative. Thus, when $x_i \in [x_i^-, x_i^+]$, the smallest possible value $\underline{s}$ of the sum

$$s \stackrel{\text{def}}{=} \sum_{i \neq k} w_{1,i} \cdot x_i$$

is attained when all x_i attain their smallest possible values $x_i = x_i^-$, and the largest possible value $\bar{s}$ of the sum s is attained when all x_i attain their largest possible values $x_i = x_i^+$:

$$\underline{s} = \sum_{i \neq k} w_{1,i} \cdot x_i^-; \quad \bar{s} = \sum_{i \neq k} w_{1,i} \cdot x_i^+.$$

Thus, we have

$$\sum_{i \neq k} w_{1,i} \cdot x_i^- \leq \sum_{i \neq k} w_{1,i} \leq \sum_{i \neq k} w_{1,i} \cdot x_i^+.$$

Now, we have two interval $[\widetilde{X}_1 - \Delta_l - w_{1,k} \cdot x_k, \widetilde{X}_1 + \Delta_l - w_{1,k} \cdot x_k]$ and $\left[\sum_{i \neq k} w_{1,i} \cdot x_i^-, \sum_{i \neq k} w_{1,i} \cdot x_i^+ \right]$ that contain the same sum $\sum_{i \neq k} w_{1,i}$. Thus, their intersection must be non-empty, i.e., the lower endpoint of the first interval cannot exceed the upper endpoint of the second interval, and vice versa (one can easily check that if these conditions are satisfied, then the above inequalities are indeed consistent):

$$\widetilde{X}_1 - \Delta_l - w_{1,k} \cdot x_k \leq \sum_{i \neq k} w_{1,i} \cdot x_i^+; \quad \sum_{i \neq k} w_{1,i} \cdot x_i^- \leq \widetilde{X}_1 + \Delta_l - w_{1,k} \cdot x_k.$$

By moving the term $w_{1,k} \cdot x_k$ to the other side of each of the inequalities and dividing both sides of each resulting inequality by a positive number $w_{1,k}$, we conclude that

$$\frac{1}{w_{1,k}} \cdot \left(\widetilde{X}_1 - \Delta_l - \sum_{i \neq k} w_{1,i} \cdot x_i^+ \right) \leq x_k \leq \frac{1}{w_{1,k}} \cdot \left(\widetilde{X}_1 + \Delta_l - \sum_{i \neq k} w_{1,i} \cdot x_i^- \right).$$

Simplest Case: Resulting Formulas. For the cells k which are not affected by the estimate $\widetilde{X}_1$, the resulting bounds on x_k are $[\underline{x}_k, \bar{x}_k]$ with $\underline{x}_k = x_k^-$ and $\bar{x}_k = x_k^+$.

For the cells k which are affected by the estimate $\widetilde{X}_1$ (i.e., for which $w_{1,k} > 0$), the resulting range $[\underline{x}_k, \bar{x}_k]$ has the form

$$\underline{x}_k = \frac{1}{w_{1,k}} \cdot \left(\widetilde{X}_1 - \Delta_l - \sum_{i \neq k} w_{1,i} \cdot x_i^+ \right); \quad \overline{x}_k = \frac{1}{w_{1,k}} \cdot \left(\widetilde{X}_1 + \Delta_l - \sum_{i \neq k} w_{1,i} \cdot x_i^- \right).$$

How to Combine Models with Different Types of Uncertainty: An Important Challenge. In the previous sections, we have described how to combine (fuse) models with probabilistic uncertainty. In this section, we have described how to combine models with interval uncertainty. But what if we need to combine models with different types of uncertainty – i.e., models with probabilistic uncertainty and models with interval uncertainty?

We may have several models with probabilistic uncertainty, and several models with interval uncertainty. In this case, we should first fuse all the models with probabilistic uncertainty into a single fused model, and fuse all the models with interval uncertainty into a single fused model. After this procedure, the original task is reduced to the task of merging two models: the first is a (combined) model with probabilistic uncertainty and the second is a (combined) model with interval uncertainty.

In general, probabilistic models provide a more detailed description of uncertainty than the interval model. Indeed, in the case of probabilistic uncertainty, we assume that we know the mean μ (equal to 0) and the standard deviation σ of the approximation error. In this case, for each certainty level p_0, we can conclude that the actual (unknown) value of the approximation error belongs to the interval $[\mu - k(p_0) \cdot \sigma, \mu + k(p_0) \cdot \sigma]$. For example, for $p_0 = 90\%$, we can take $k(p_0) = 2$; for $p_0 = 99.9\%$, we can take $k(p_0) = 3$, etc.

In case of interval uncertainty, we only know the interval $[-\Delta, \Delta]$ of possible values of approximation error. In this case, we do not know the exact values of μ and σ, we can only conclude that the actual (unknown) values of μ and σ satisfy the conditions $-\Delta \leq \mu - k(p_0) \cdot \sigma$ and $\mu + k(p_0) \cdot \sigma \leq \Delta$. In other words, the second (interval) uncertainty model corresponds to the whole class of possible probabilistic uncertainty models. So, a natural way to combine the probabilistic and the interval models is to consider the combinations of the first probabilistic model with all possible probabilistic models corresponding to the second (interval) model.

For example, as we have mentioned earlier, if we fuse n values $\widetilde{x}^{(i)}$ whose measurement errors are random with mean 0 and known standard deviations $\sigma^{(i)}$, then, as a result of a fusion, we get an estimate $x = \dfrac{\sum\limits_{i=1}^{n} \widetilde{x}^{(i)} \cdot (\sigma^{(i)})^{-2}}{\sum\limits_{i=1}^{n} (\sigma^{(i)})^{-2}}$ whose standard deviation is equal to $\sigma^{-2} = \sum\limits_{i=1}^{n} (\sigma^{(i)})^{-2}$. If we only know, e.g., the n-value $x^{(n)}$ with interval uncertainty, i.e., if we only know the bounds $\underline{x}^{(n)}$ and $\overline{x}^{(n)}$ for which $\underline{x}^{(n)} \leq x^{(n)} \leq \overline{x}^{(n)}$, then, in contrast to the probabilistic case, we do not know the exact mean $\widetilde{x}^{(n)}$ and standard deviation $\sigma^{(n)}$ corresponding to the n-th measurement; instead, we only know that, for an appropriately chosen $k_0 = k(p_0)$, we have $\underline{x}^{(n)} \leq \widetilde{x}^{(n)} - k_0 \cdot \sigma^{(n)}$ and $\widetilde{x}^{(n)} + k_0 \cdot \sigma^{(n)} \leq \overline{x}^{(n)}$. Thus, for the fused result, instead of a single value x, we now have a whole range of values, namely, the set of all possible

values of the ratio $x = \dfrac{\sum\limits_{i=1}^{n-1} \tilde{x}^{(i)} \cdot (\sigma^{(i)})^{-2} + \tilde{x}^{(n)} \cdot (\sigma^{(n)})^{-2}}{\sum\limits_{i=1}^{n-1} (\sigma^{(i)})^{-2} + (\sigma^{(n)})^{-2}}$ corresponding to all pos-

sible values $\tilde{x}^{(n)}$ and $\sigma^{(n)}$ that satisfy the above two inequalities. Similarly, we can handle the cases when we have more data points known with interval uncertainty, and the cases when we also need to take into account spatial resolution.

As one can see from this description, even in the simplest case, to combine probabilistic and interval uncertainty, we need to solve a complex non-linear optimization problem. Thus, combining interval and probabilistic uncertainty remains an important computational challenge.

4.4 Fuzzy Techniques: Imprecise (Fuzzy) Uncertainty Leads to More Computationally Efficient Processing

As we mentioned in Chapter 3, in addition to probabilistic and interval uncertainty, we also encounter imprecise(fuzzy) uncertainty. In this section, on the example of using machine learning to diagnose cancer based on the microarray gene expression data, we show that fuzzy-technique description of imprecise knowledge can improve the efficiency of the existing algorithms. Specifically, we show that the fuzzy-technique description leads to a formulation of the learning problem as a problem of sparse optimization, and we can therefore use computationally efficient ℓ_1-techniques to solve the resulting optimization problem.

The results of this section first appeared in [64].

Machine Learning: A Typical Problem. In machine learning, we know how to classify several known objects, and we want to learn how to classify new objects.

For example, in a biomedical application, we have microarray data corresponding to healthy cells and microarray data corresponding to different types of tumors. Based on these samples, we would like to be able, given a microarray data, to decide whether we are dealing with a healthy tissue or with a tumor, and if it is a tumor, what type of cancer does the patient have.

In general, each object is characterized by the results

$$x = (x_1, \ldots, x_n)$$

of measuring several (n) different quantities. So, in mathematical terms, machine learning can be described as a following problem (see, e.g., [5]):

- we have K possible labels $1, \ldots, K$ describing different classes;
- we have several vectors $x(j) \in R^n$, $j = 1, \ldots, N$, each of which is labeled by an integer $k(j)$ ranging from 1 to K; vectors labeled as belonging to the k-th class will be also denoted by

$$x(k,1),\ldots,x(k,N_k);$$

- we want to use these vectors to assign, to each new vector $x \in R^n$, a value

$$k \in \{1,\ldots,K\}.$$

Machine Learning: Original Idea. The first machine learning algorithms were based on the assumption that for each class C_k, the set of the vectors x corresponding to this class is *convex*, i.e., that for every two vectors $x, x' \in C_k$, and for every number $\alpha \in (0,1)$, their *convex combination*

$$\alpha \cdot x + (1 - \alpha) \cdot x'$$

also belongs to the class C_k.

It is known that if different classes $C_1,\ldots,C_K$ are convex, then we can separate them by using linear separators; see, e.g., [62]. For example, if we only have two classes C_1 and C_2, then there exists a linear function

$$f(x) = c_0 + \sum_{i=1}^{n} c_i \cdot x_i$$

and a threshold value y_0 such that:

- for all vectors $x \in C_1$, we have $f(x) < y_0$, while
- for all vectors $x \in C_2$, we have $f(x) > y_0$.

Thus, if we know that an object characterized by a vector x belongs to one of these two classes, and we want to decide to which of these classes it belongs, we can compute the value of the linear function $f(x)$, and then:

- if $f(x) < y_0$, conclude that x belongs to the class C_1, and
- if $f(x) > y_0$, conclude that x belongs to the class C_2.

If we have more than two classes, then, to classify an object, we can use linear functions separating different pairs of classes.

Machine Learning: Current Development. In practice, the classes C_k are not convex. As a result, it is often impossible to use simple linear separators $f(x)$ to separate different classes, we need *nonlinear* separating functions.

The first such separating functions came from an observation that in the biological neurons – cells that enable us to classify objects – the dependence of the output on the input is nonlinear. The resulting Artificial Neural Networks simulating this nonlinearity turned out to be an efficient machine learning tool.

Even more efficient algorithms became possible when researchers realized that a general nonlinear separating function $f(x_1,\ldots,x_n)$ can be represented, e.g., by its Taylor series

$$f(x_1,\ldots,x_n) = c_0 + \sum_{i=1}^{n} c_i \cdot x_i + \sum_{i=1}^{n}\sum_{j=1}^{n} c_{ij} \cdot x_i \cdot x_j + \ldots$$

This expression becomes linear if, in addition to the original values $x_1, \ldots, x_n$, we also add their combinations such as $x_i \cdot x_j$. The corresponding *Support Vector Machine* (SVM) techniques [5, 75] are, at present, the most efficient in machine learning. For example, SVM is implemented in the GEMS-SVM software which is used to automatically diagnose cancer based on the microarray gene expression data [74].

Remaining Problem. While the SVM methods lead to efficient classification results, these results are not perfect. If we:

- divide the original samples into a training set and a training set,
- train an SVM method on the training set, and then
- test the resulting classification on a testing set,

then we get, depending on the type of tumor, 90 to 100% correct classifications. 90% is impressive, but it still means that up to 10% of all the patients are misclassified. How can we improve this classification?

Our Idea. As we have mentioned, linear algorithms are based on an assumption that all the classes C_k are convex. Since these classes are not exactly convex, SVM techniques abandon the efficient linear separation algorithms and use less efficient general nonlinear techniques.

In reality, while the classes are *not exactly convex*, they are *somewhat* convex, in the sense that for many vectors x and x' from each class C_k and for many values α, the convex combination $\alpha \cdot x + (1 - \alpha) \cdot x'$ still belongs to C_k.

In this section, we use fuzzy techniques to formalize this imprecise idea of "somewhat" convexity, and we show that the resulting machine learning algorithm indeed improves the efficiency.

Need to Use Degrees. The usual ("crisp") description of convexity is that if the two vectors x, x' belong to the set C_k of all the vectors corresponding to the k-th class, then their convex combination

$$\alpha \cdot x + (1 - \alpha) \cdot x'$$

also belongs to this class C_k.

"Somewhat" convexity means that if $x, x' \in C_k$, then we can only conclude with a certain *degree of confidence* that the object corresponding to the vector

$$\alpha \cdot x + (1 - \alpha) \cdot x'$$

belongs to the k-th class. So, to get an adequate description of our knowledge about the k-th class, we need to assign, to each possible vector x, a degree $\mu_k(x)$ to which we are confident that an object corresponding to this vector x belongs to this class. The corresponding function $x \rightarrow \mu_k(x) \in [0, 1]$ constitutes a *fuzzy set* [31, 50, 78].

For each vector x, we can then assign the corresponding object to the class k for which we are most confident that x belongs to this class, i.e., to the class k for which the degree of confidence $\mu_k(x)$ is the largest possible.

Using "Somewhat" Convexity to Estimate Degrees of Confidence. For each class k, we know that several "sample" vectors that for sure belong to this class – namely, the vectors

$$x(k,1),\ldots,x(k,N_k)$$

which have been originally labeled as belonging to the j-th class.

We also know that a convex combination

$$x'' \stackrel{\text{def}}{=} \alpha \cdot x + (1-\alpha) \cdot x'$$

of two vectors x and x' belongs to the k-th class if both x and x' belong to the class, and if the convexity rule holds for this situation. Thus, the degree $\mu_k(x'')$ to which x'' belongs to the k-th class is greater than or equal to the degree

$$\mu((x \in C_k) \,\&\, (x' \in C_k) \,\&\, \text{rule holds})$$

with which all these three conditions hold. A simple way to estimate the degree to which all three conditions hold is to take a product of the degree of the component statements:

- the product corresponds to the case when degrees are (subjective) probabilities and all three events are reasonably independent; see, e.g., [71];
- the product is also one of the simplest and most widely used ways of combining degrees of confidence in fuzzy applications.

So, if we denote, by r, our degree of confidence in the convexity rule, we conclude that

$$\mu_k(\alpha \cdot x + (1-\alpha) \cdot x') \geq r \cdot \mu_k(x) \cdot \mu_k(x').$$

This means that for vectors x'' which are convex combinations of two sample vectors, the degree $\mu_k(x'')$ is great than or equal to r.

A vector x''' which is a convex combinations of *three* sample vectors x, x', and x'' can be represented as a convex combination of x and a convex combination y of x' and x''; since $\mu_k(y) \geq r$, we thus get

$$\mu_k(x''') \geq r \cdot \mu_k(x) \cdot \mu_k(y) \geq r \cdot 1 \cdot r = r^2.$$

Similarly, for a vector y which is a convex combination of four sample vectors, we get $\mu_k(y) \geq r^3$. In general, for a vector y which is a combination of v sample vectors, we get $\mu_k(y) \geq r^{v-1}$. In other words, for a vector

$$y = \sum_{j=1}^{N_k} \alpha_j \cdot x(k,j),$$

where $\alpha_j \geq 0$ and $\sum\limits_{j=1}^{N_k} \alpha_j = 1$, we conclude that

$$\mu_k(y) \geq r^{\|\alpha\|_0 - 1},$$

where the ℓ_0-norm $\|\alpha\|_0$ of a vector $\alpha = (\alpha_1, \ldots, \alpha_{N_k})$ is defined as the number of non-zero components of this vector:

$$\|\alpha\| \stackrel{\text{def}}{=} \#\{i : \alpha_j \neq 0\}.$$

Using Closeness. Another natural idea is that if a vector x is *close* to a vector y corresponding to the object from the k-th class, then it is reasonable to conclude (with some degree of confidence) that the object corresponding to x also belongs to the same class. The smaller the distance

$$\|x - y\|_2 \stackrel{\text{def}}{=} \sqrt{\sum_{i=1}^{n} (x_i - y_i)^2}$$

between the vectors x and y, the larger our degree of certainty that x belongs to this class.

In this paper, we assume that this degree of certainty is described by a Gaussian expression $\exp\left(-\dfrac{\|x - y\|_2^2}{\sigma^2}\right)$ for an appropriate value σ. Similarly to selecting a product, we selected the Gaussian formula for two reasons:

- the Gaussian distribution is ubiquitous in probability theory, since due to the Central Limit Theorem, every time when the effect is caused by a large number of small independent factors, the distribution is close to Gaussian; see, e.g., [71];
- in fuzzy logic, Gaussian membership functions have been successfully used in many practical applications to describe closeness.

As a result, for every two vectors x and y, we have

$$\mu_k(x) \geq \mu_k(y) \cdot \exp\left(-\frac{\|x - y\|_2^2}{\sigma^2}\right).$$

Resulting Formula for the Degree of Confidence. By combining the two formulas describing "somewhat" convexity and closeness, we conclude that for every vector x and for every vector α, we have

$$\mu_k(x) \geq \exp\left(-\frac{\left\|x - \sum\limits_{j=1}^{N_k} \alpha_j \cdot x(k,j)\right\|_2^2}{\sigma^2}\right) \cdot r^{\|\alpha\|_0 - 1}.$$

This is true for every α, so we can conclude that the desired degree of confidence is larger than or equal than the largest of the right-hand sides:

$$\mu_k(x) \geq \max_{\alpha} \exp\left(-\frac{\left\| x - \sum_{j=1}^{N_k} \alpha_j \cdot x(k,j) \right\|_2^2}{\sigma^2} \right) \cdot r^{\|\alpha\|_0 - 1}.$$

This right-hand side is all we can state about the vector x, so we can thus conclude that

$$\mu_k(x) = \max_{\alpha} \exp\left(-\frac{\left\| x - \sum_{j=1}^{N_k} \alpha_j \cdot x(k,j) \right\|_2^2}{\sigma^2} \right) \cdot r^{\|\alpha\|_0 - 1}.$$

Thus, to classify a vector x, we need to compute the values of the right-hand side corresponding to different classes k, and select the class k for which the value $\mu_k(x)$ is the largest.

Formula Simplified. Maximizing the value $\mu_k(x)$ is equivalent to minimizing a simpler expression

$$d_k(x) \stackrel{\text{def}}{=} -\ln(\mu_k(x))$$

which can be described as follows:

$$d_k(x) = \min_{\alpha} \left(\frac{\left\| x - \sum_{j=1}^{N_k} \alpha_j \cdot x(k,j) \right\|_2^2}{\sigma^2} + (\|\alpha\|_0 - 1) \cdot R \right),$$

where $R \stackrel{\text{def}}{=} -\ln(r)$. To use this formula, for each k, we need to minimize the expression

$$\frac{\left\| x - \sum_{j=1}^{N_k} \alpha_j \cdot x(k,j) \right\|_2^2}{\sigma^2} + (\|\alpha\|_0 - 1) \cdot R$$

with respect to vectors α. By adding a constant R to this expression and dividing the resulting objective function by R, we get an equivalent problem of optimizing the objective function

$$\mathscr{C} \cdot \left\| x - \sum_{j=1}^{N_k} \alpha_j \cdot x(k,j) \right\|_2^2 + \|\alpha\|_0,$$

where

$$\mathscr{C} \stackrel{\text{def}}{=} \frac{1}{R \cdot \sigma^2}.$$

The above Optimization Problem Is Hard to Solve Exactly. The main algorithmic problem with the above optimization is that optimization problems involving the ℓ_0-norm $\|\alpha\|_0$ are, in general, NP-hard; see, e.g., [11]. This means, crudely speaking, that unless P = NP (which most computer scientists believe to be impossible), there is no hope to have an efficient algorithm for exactly solving all particular cases of such an optimization problem.

ℓ_1-Optimization as a Good Approximation to ℓ_0-Optimization Problems. In terms of the Lagrange multiplier method, the above optimization problem is equivalent to minimizing the ℓ_0-norm $\|\alpha\|_0$ under the constraint

$$\left\| x - \sum_{j=1}^{N_k} \alpha_j \cdot x(k, j) \right\|_2 \leq C,$$

for an appropriate constant C. One of the reasons why this problem is NP-hard is that while the set of all the vectors α which satisfy this constraint is convex, the ℓ_0-norm $\|\alpha\|_0$ is not convex. To make the problem computationally efficient, researchers proposed to approximate it by a convex objective function, namely, approximate it by an ℓ_1-norm

$$\|\alpha\|_1 \overset{\text{def}}{=} \sum_{j=1}^{N_k} |\alpha_j|.$$

It turns out that not only we have a good approximation in many practical applications, but in many cases, ℓ_1-optimization is *equivalent* to ℓ_0-optimization; see, e.g., [8].

We therefore propose to use the replace each original NP-hard optimization problem with a feasible convex optimization problem of minimizing

$$\mathscr{C}' \cdot \left\| x - \sum_{j=1}^{N_k} \alpha_j \cdot x(k, j) \right\|_2^2 + \|\alpha\|_1$$

for an appropriate constant $\mathscr{C}'$.

Taking the Specific Problem into Account. The above formulation works for any "somewhat" convex case.

For microarray analysis, there is a specific property – that the actual values of the vector x depend on the efficiency of the microarray technique. In other words, with a less efficient technique, we will get $\lambda \cdot x$ for some constant λ. From this viewpoint, it is reasonable to use not just *convex* combinations, but arbitrary *linear* combinations of the original vectors $x(k, j)$.

Towards Even More Efficient Computations. While ℓ_1-optimization is efficient, it still takes a large amount of computation times.

In our formulation, we need to repeat this optimization as many times as there are classes; this repetition further increases the computation time.

To decrease computation time, we propose the following idea:

- instead of trying to represent the vector x as a linear combination of vectors from each class,
- let us look for a representation of x as a linear combination of *all* sample vectors, from all classes.

In other words, instead of solving K different optimization problems, let us solve a single problem of minimizing the expression

$$\mathscr{C}' \cdot \left\| x - \sum_{j=1}^{N} \alpha_j \cdot x(j) \right\|_2^2 + \|\alpha\|_1.$$

Then, for each class k, we only take the components belonging to this class, and select a class for which the resulting linear combination is the closest to the original vector x, i.e., for which the easy-to-compute distance

$$\left\| x - \sum_{j:k(j)=k} \alpha_j \cdot x(j) \right\|_2$$

is the smallest possible.

Observation. Interestingly, this time-saving idea not only increased the efficiency of our method, it also improve the quality of classification.

We think that this improvement is related to the fact that all the data contain measurement noise. On each computation step, we process noisy data and the results combine noise from different inputs of this step – and thus, get noisier and noisier with each computation step. From this viewpoint, the longer computations, the more noise we add.

So, while our approximate method does not lead to a perfect minimum in the original optimization problem, it saves on the noise, and these savings overwhelm the errors introduced by replacing the original K times repeated optimizations with a single "approximate" ones.

Results. Detailed results are presented in [63]. (It should be mentioned that the paper [63] does not include any justification of our proposed algorithm; such a justification is presented here for the first time.) The results are good; for example:

- for brain tumor, the probability of correct identification increased from 90% for the best known SVM techniques to 91% for our method;
- for prostate tumor, the probability similarly increased from 93% to 94%.

In some cases, the success probability of our algorithm is slightly lower than for the SVM techniques. However, it should be mentioned that the good SVM results are partly due to the fact that to make SVM efficient, we need to select appropriate non-linear functions, and there are known algorithms for selecting these functions. If we select arbitrary functions, we usually get not-so-good results; this selection must be expertly tuned to the problem. Hence, the very fact that SVM results (obtained after trying several different sets of nonlinear functions are selecting the most appropriate set) are good does not necessarily mean that we will get similarly good results in other cases.

In contrast, our sparse method has only one parameter to tune – the parameter $\mathscr{C}'$. From this viewpoint, our technique is much less subjective, much more reliable – and leads to similar or even better classification results.

4.5   Beyond Algorithmic Approaches to Computational Efficiency

Even when all reasonable algorithmic speed-up ideas have been implemented, the computation time is still often too long. In such situations, the only remaining way to speed up computations is to use different hardware speed-up ideas. Such ideas range from currently available (like parallelization) to more futuristic ideas like quantum computing. While parallelization has been largely well-researched, the use of future techniques (such as quantum computing) in data processing and uncertainty estimation is still largely an open problem. In this section, we show how quantum computing can be used to speed up the corresponding computations.

Case Study: Reliability of Interval Data. In interval computations, i.e., in processing interval data, we usually assume that all the measuring instruments functioned correctly, and that all the resulting intervals

$$[\widetilde{x} - \Delta, \widetilde{x} + \Delta]$$

indeed contain the actual value x.

In practice, nothing is 100% reliable. There is a certain probability that a measurement instrument malfunctions. As a result, when we repeatedly measure the same quantity several times, we may have a certain number of measurement results (and hence intervals) which are "way off", i.e., which do not contain the actual value at all.

For example, when we measure the temperature, we will usually get values which are close to the actual temperature, but once in a while the thermometer will not catch the temperature at all, and return a meaningless value like 0. It may be the fault of a sensor, and/or it may be a fault of the processor which processes data from the sensor. Such situations are rare, but when we process a large amount of data, it is typical to encounter some outliers.

Such outliers can ruin the results of data processing. For example, if we compute the average temperature in a given geographic area, then averaging the correct measurement results would lead a good estimate, but if we add an outlier, we can get a nonsense result. For example, based on the measurements of temperature in El Paso in Summer resulting in 95, 100, and 105, we can get a meaningful value

$$\frac{95 + 100 + 105}{3} = 100.$$

However, if we add an outlier 0 to this set of data points, we get a misleading estimate

$$\frac{95 + 100 + 105 + 0}{4} = 75$$

creating the false impression of El Paso climate.

A natural way to characterize the reliability of the data is to set up a bound on the probability p of such outliers. Once we know the value p, then, out of n results of measuring the same quantity, we can dismiss $k \overset{\text{def}}{=} p \cdot n$ largest values and k smallest values, and thus make sure that the outliers do not ruin the results of data processing.

Need to Gauge the Reliability of Interval Data. Where does the estimate p for data reliability come from? The main idea of gauging this value comes from the fact that if we measure the same quantity several times, and all measurements are correct (no outliers), then all resulting intervals $\mathbf{x}^{(1)}, \ldots, \mathbf{x}^{(n)}$ contain the same (unknown) value x – and thus, their intersection is non-empty.

If we have an outlier, then it is highly probably that this outlier will be far away from the actual value x – and thus, the intersection of the resulting n intervals (including intervals coming from outliers) will be empty.

In general, if the percentage of outliers does not exceed p, then we expect that out of n given intervals, at least $n - k$ of these intervals (where $k \overset{\text{def}}{=} p \cdot n$) correspond to correct measurements and thus, have a non-empty intersection.

So, to check whether our estimate p for reliability is correct, we must be able to check whether out of the set of n given intervals, there exists a subset of $n - k$ intervals which has a non-empty intersection.

Need to Gauge Reliability of Interval Data: Multi-D Case. In the previous section, we considered a simplified situation in which each measuring instrument measures exactly one quantity. In practice, a measuring instrument often measure several different quantities $x_1, \ldots, x_d$. Due to uncertainty, after the measurement, for each quantity x_i, we have an interval $\mathbf{x}_i$ of possible values. Thus, the set of all possible values of the tuple $x = (x_1, \ldots, x_d)$ is a *box*

$$X = \mathbf{x}_1 \times \ldots \times \mathbf{x}_d = \{(x_1, \ldots, x_d) : x_1 \in \mathbf{x}_1, \ldots, x_d \in \mathbf{x}_d\}.$$

In this multi-D case, if all the measurements are correct (no outliers), all the corresponding boxes $X^{(1)}, \ldots, X^{(n)}$ contain the actual (unknown) tuple and thus, the intersection of all these boxes is non-empty.

Thus, to check whether our estimate p for reliability is correct, we must be able to check whether out of the set of n given boxes, there exists a subset of $n - k$ boxes which has a non-empty intersection.

Resulting Computational Problem: Box Intersection Problem. Thus, both in the interval and in the fuzzy cases, we need to solve the following computational problem:

Given:

- integers d, n, and k; and
- n d-dimensional boxes $X^{(j)} = [\underline{x}_1^{(j)}, \overline{x}_1^{(j)}] \times \ldots \times [\underline{x}_n^{(j)}, \overline{x}_n^{(j)}]$, $j = 1, \ldots, n$, with rational bounds $\underline{x}_i^{(j)}$ and $\overline{x}_i^{(j)}$.

Check: whether we can select $n - k$ of these n boxes in such a way that the selected boxes have a non-empty intersection.

First Result: The Box Intersection Problem Is NP-Complete. The first result related to this problem is that in general, the above box intersection problem is NP-complete.

The proof of this result is given at the end of this section.

The Meaning of NP-Completeness: A Brief Explanation. Crudely speaking, NP-completeness means that it is impossible to have an efficient algorithm that solves all particular instances of the above computational problem.

The notion of NP-completeness is relayed to the fact that some algorithms require so much computation time that even for inputs of reasonable size, the required computation time exceeds the lifetime of the Universe – and thus, cannot be practically computed. For example, if for n inputs, the algorithm requires 2^n units of time, then for $n \approx 300 - 400$, the resulting computation time is un-realistically large. How can we separate "realistic" ("feasible") algorithms from non-feasible ones?

The running time of an algorithm depends on the size of the input. In the computer, every object is represented as a sequence of bits (0s and 1s). Thus, for every computer-represented object x, it is reasonable to define its *size* (or *length*) $\text{len}(x)$ as the number of bits in this object's computer representation.

It is known that in most feasible algorithms, the running time on an input x is bounded either by the size of the input, or by the square of the size of the input, or, more generally, by a polynomial of the size of the input. It is also known that in most non-feasible algorithms, the running time grows exponentially (or even faster) with the size, so it cannot be bounded by any polynomial. In view of this fact, in theory of computation, an algorithm is usually called feasible if its running time is bounded by a polynomial of the size of the input. This definition is not perfect: e.g., if the running time on input of size n is $10^{40} \cdot n$, then this running time is bounded by a polynomial but it is clearly not feasible. However, this definition is the closest to the intuitive notion of feasible, and thus, the best we have so far.

According to this definition, an algorithm A is called *polynomial time* if there exists a polynomial $P(n)$ such that on every input x, the running time of the algorithm A does not exceed $P(\text{len}(x))$. The class of all the problems which can be solved by polynomial-time algorithms is denoted by P.

What do we mean by "a problem"? In most practical situations, to solve a problem means to find a solution that satisfies some (relatively) easy-to-check constraint: e.g., to design a bridge that can withstand a certain amount of load and wind, to design a spaceship and its trajectory that enables us to deliver a robotic rover to Mars, etc. In all these cases, once we have a candidate for a solution, we can check, in reasonable (polynomial) time whether this candidate is indeed a solution. In other words, once we guessed a solution, we can check its correctness in polynomial time. In theory of computation, this procedure of guess-then-compute is called *non-deterministic computation*, so the class of all problems whose solution can be checked in polynomial time is called Non-deterministic Polynomial, or NP, for short.

Most computer scientists believe that not all problems from the class NP can be solved in polynomial time, i.e., that $NP \neq P$. However, no one has so far been able to prove that this belief is indeed true. What is known is that some problems from the class NP are the hardest in this class – in the sense that every other problem from the class NP can be reduced to such a problem.

Specifically, a general problem (not necessarily from the class NP) is called *NP-hard* if every problem from the class NP can be reduced to particular cases of this problem. If a problem from the class NP is NP-hard, we say that it is *NP-complete*.

One of the best known examples of NP-complete problems is the problem of *propositional satisfiability* for formulas in 3-Conjunctive Normal Form (3-CNF). Let us describe this problem is some detail. We start with v Boolean variables $z_1, \ldots, z_v$, i.e., variables which can take only values "true" or "false". A *literal* ℓ is defined as a variable z_i or its negation $\neg z_i$. A *clause* is defined as a formula of the type $\ell_1 \vee \ell_2 \vee \ldots \vee \ell_m$. Finally, a *propositional formula in Conjunctive Normal Form (CNF)* is defined as a formula F of the type $C_1 \& \ldots \& C_n$, where $C_1, \ldots, C_n$ are clauses. This formula is called a *3-CNF* formula if every clause has at most 3 literals, and a *2-CNF* formula if every clause has at most 2 literals.

The propositional satisfiability problem is as follows:

- Given a propositional formula F (e.g., a formula in CNF);
- Determine if there exist values of the variables $z_1, \ldots, z_v$ which make the formula F true.

For the propositional satisfiability problem, the proof of NP-hardness is somewhat complex. However, once this NP-hardness is proven, we can prove the NP-hardness of other problems by reducing satisfiability to these problems.

Indeed, by definition, NP-hardness of satisfiability means that every problem from the class NP can be reduced to satisfiability. If we can reduce satisfiability to some other problem, this means that by combining these two reductions, we can reduce every problem from the class NP to this new problem – and thus, that this new problem is also NP-hard.

For a more detailed and more formal definition of NP-hardness, see, e.g., [34, 57].

Case of Fixed Dimension: Efficient Algorithm for Gauging Reliability. In general, when we allow unlimited dimension d, the box intersection problem (computational problem related to gauging reliability) is computationally difficult (NP-hard).

In practice, however, the number d of quantities measured by a sensor is small: e.g.,

- a GPS sensor measures 3 spatial coordinates;
- a weather sensor measures (at most) 5 quantities: temperature, atmospheric pressure, and the 3 dimensions of the wind vector.

It turns out that if we limit ourselves to the case of a fixed dimension d, then we can solve the above computational problem in polynomial time $O(n^d)$; see, e.g., [15].

Indeed, for each of d dimensions x_i ($1 \le i \le d$), the corresponding n intervals have $2n$ endpoints $\underline{x}_i^{(j)}$ and $\overline{x}_i^{(j)}$. Let us show that if there exists a vector x which belongs to $\ge n - k$ boxes $X^{(j)}$, then there also exists another point y with this property in which every coordinate y_i coincides with one of the endpoints. Indeed, if for some i, the value x_i is not an endpoint, then we can take the closest endpoint as y_i. One can easily check that this change will keep the vector in all the boxes $X^{(j)}$.

Thus, to check whether there exists a vector x that belongs to at least $n - k$ boxes $X^{(j)}$, it is sufficient to check whether there exist a vector formed by endpoints which satisfies this property. For each vector $y = (y_1, \ldots, y_d)$ and for each box $X^{(j)}$, it takes $d = O(1)$ steps to check whether $y \in X^{(j)}$. After repeating this check for all n boxes, we thus check whether this vector y satisfies the desired property in time $n \cdot O(1) = O(n)$.

For each of d dimensions, there are $2n$ possible endpoints; thus, there are $(2n)^d$ possible vectors y formed by such endpoints. For each of these vectors, we need time $O(n)$, so the overall computation time for this procedure requires time $O(n) \cdot (2n)^d = O(n^{d+1})$ – i.e., indeed time which grows polynomially with n.

Remaining Problem. In the previous section, we have shown that for a bounded dimension d, we can solve the box intersection problem in polynomial time. However, as we have mentioned, polynomial time does not always mean that the algorithm is practically feasible.

For example, for a meteorological sensor, the dimension d is equal to 5, so we need n^6 computational steps. For $n = 10$, we get 10^6 steps, which is easy to perform. For $n = 100$, we need $100^6 = 10^{12}$ steps which is also doable – especially on a fast computer. However, for a very reasonable amount of $n = 10^3 = 1000$ data points, the above algorithm requires $1000^6 = 10^{18}$ computational steps – which already requires a long time, and for $n = 10^4$ data points, the algorithm requires a currently practically impossible amount of 10^{24} computational steps.

It is therefore desirable to speed up the computations. Here, even when all reasonable algorithmic speed-up ideas have been implemented, the computation time is still often too long. In such situations, the only remaining way to speed up computations is to use different hardware speed-up ideas. Such ideas range from currently available (like parallelization) to more futuristic ideas like quantum computing. While parallelization has been largely well-researched, the use of future techniques (such as quantum computing) in data processing and uncertainty estimation is still largely an open problem.

In this section, we show that in our problem, we can indeed achieve a significant speed up if we use quantum computations.

Quantum Computations: A Reminder. Before we explain how exactly quantum computations can speed up the computations needed to gauge reliability, let us briefly recall how quantum effects can be used to speed up computations.

In this chapter, we will use Grover's algorithm for quantum search. Without using quantum effects, we need – in the worst case – at least N computational steps to search for a desired element in an unsorted list of size N. A quantum computing algorithm proposed by Grover (see, e.g., [16, 17, 51]) can find this element much faster – in $O(\sqrt{N})$ time.

Specifically, Grover's algorithm, given:

- a database $a_1, \ldots, a_N$ with N entries,
- a property P (i.e., an algorithm that checks whether P is true), and
- an allowable error probability δ,

returns, with probability $\geq 1 - \delta$, either the element a_i that satisfies the property P or the message that there is no such element in the database.

This algorithm requires $c \cdot \sqrt{N}$ steps (= calls to P), where the factor c depends on δ (the smaller δ we want, the larger c we must take).

For the Grover's algorithm, the entries a_i do not need to be all physically given, it is sufficient to have a procedure that, given i, produces a_i.

Brassard et al. used the ideas behind Grover's algorithm to produce a new quantum algorithm for *quantum counting*; see, e.g., [7, 51]. Their algorithm, given:

- a database $a_1, \ldots, a_N$ with N entries,
- a property P (i.e., an algorithm that checks whether P is true), and
- an allowable error probability δ,

returns an approximation $\tilde{t}$ to the total number t of entries a_i that satisfy the property P.

This algorithm contains a parameter M that determines how accurate the estimates are. The accuracy of this estimate is characterized by the inequality

$$\left| \tilde{t} - t \right| \leq \frac{2\pi}{M} \cdot \sqrt{t} + \frac{\pi^2}{M^2} \tag{4.1}$$

that is true with probability $\geq 1 - \delta$.

This algorithm requires $c \cdot M \cdot \sqrt{N}$ steps (= calls to P), where the factor c depends on δ (the smaller δ we want, the larger c we must take).

In particular, to get the exact value t, we must attain accuracy $\left| \tilde{t} - t \right| \leq 1$, for which we need $M \approx \sqrt{N}$. In this case, the algorithm requires $O(\sqrt{t \cdot N})$ steps.

Quantum Computations Can Drastically Speed Up Gauging Reliability. As a part of the above algorithm for checking box intersections, we search among $O(n^d)$ vectors y for a vector that belongs to at least $n - k$ boxes $X^{(j)}$. For each of these vectors y, we need to find to how many of n boxes $X^{(j)}$ the vector y belongs; this requires time $O(n)$.

For each vector y, we can use the quantum counting algorithm to compute the number of boxes in time $O(\sqrt{n})$. We can then use Grover's algorithm to reduce the non-quantum search of $N = O(n^d)$ vectors to a search whose time is equivalent to processing $\sqrt{N} = O(n^{d/2})$ such vectors. For each of these vectors, we need time

$O(\sqrt{n})$. Thus, if we use quantum computations, we need the total computation time $O(n^{d/2}) \cdot O(\sqrt{n}) = O(n^{(d+1)/2})$.

This time is much smaller than the non-quantum computation time $O(n^{d+1})$. For example, for the above meteorological example of $n = 10^4$ and $d = 5$, the non-quantum algorithm requires a currently impossible amount of 10^{24} computational steps, while the quantum algorithm requires only a reasonable amount of 10^{12} steps.

Comment. A similar square root reduction can be achieved in the general case, but for general d, $n^{(d+1)/2}$ computational steps may still take too long.

Conclusion. In traditional interval computations, we assume that the interval data corresponds to guaranteed interval bounds, and that fuzzy estimates provided by experts are correct. In practice, measuring instruments are not 100% reliable, and experts are not 100% reliable, we may have estimates which are "way off", intervals which do not contain the actual values at all. Usually, we know the percentage of such outlier un-reliable measurements. It is desirable to check that the reliability of the actual data is indeed within the given percentage. In this section, we have shown that:

- in general, the problem of checking (gauging) this reliability is computationally intractable (NP-hard);
- in the reasonable case when each sensor measures a small number of different quantities, it is possible to solve this problem in polynomial time; and
- quantum computations can drastically reduce the required computation time.

Proof That the Box Intersection Problem Is NP-Hard. As we have mentioned in the main text, in gauging reliability, it is important to be able to solve the following box intersection problem:

- Given: a set of n d-dimensional boxes, and a number $k < n$.
- Check: is there a vector x which belongs to at least $n - k$ of these n boxes?

This box intersection problem obviously in NP: it is easy to check that a given vector x belongs to each of the boxes, and thus, to check whether it belongs to at least $n - k$ of the boxes. So we only need a proof of NP-hardness.

The proof is by reduction from the following auxiliary "limited clauses" problem which has been proved to be NP-complete:

- Given: a 2-CNF formula F and a number k,
- check: is there a Boolean vector which satisfies at most k clauses of F.

This problem was proved to be NP-complete in [32] (see also [2], p. 456).

As we have mentioned in the main text of this chapter, to prove the NP-hardness of our box intersection problem, it is therefore sufficient to be able to reduce this "limited clauses" problem to the box intersection problem.

Indeed, suppose that we are given a 2-CNF formula F. Let us denote the number of Boolean variables in this formula by d, and the overall number of clauses in this formula F by n. Based on the formula F, let us build a set of n d-dimensional boxes, one for each clause. If clause C_i contains Boolean variables z_{i1} and z_{i2}, then the i-th box $X^{(i)}$ has sides $[0,1]$ in all dimensions except in the dimensions associated with variables z_{i1} and z_{i2}. For those two dimensions, the side is:

- $[0,0]$ if the variable occurs positively in the clause
  (i.e., if the clause contains the positive literal z_{ij}), and
- $[1,1]$ is the variable occurs negatively in the clause
  (i.e., if the clause contains the negative literal $\neg z_{ij}$).

According to the construction:

- for a clause $z_{i1} \vee z_{i2}$, a vector x belongs to the box

$$X^{(i)} = \ldots \times [0,1] \times [0,0] \times [0,1] \times \ldots \times [0,1] \times [0,0] \times [0,1] \times \ldots$$

  if and only of $x_{i1} = 0$ and $x_{i2} = 0$;
- for a clause $z_{i1} \vee \neg z_{i2}$, a vector x belongs to the box $X^{(i)}$
  if and only of $x_{i1} = 0$ and $x_{i2} = 1$;
- for a clause $\neg z_{i1} \vee z_{i2}$, a vector x belongs to the box $X^{(i)}$
  if and only of $x_{i1} = 1$ and $x_{i2} = 0$;
- for a clause $\neg z_{i1} \vee \neg z_{i2}$, a vector x belongs to the box $X^{(i)}$
  if and only of $x_{i1} = 1$ and $x_{i2} = 1$.

The claim is that there exists a vector x which belongs to at least $n - k$ of these n boxes if and only if there is a Boolean vector z which satisfies at most k clauses of the formula F.

Suppose that there exists a vector x which belongs to at least $n - k$ of these n boxes. According to our construction, each box $X^{(i)}$ comes from a clause C_i that contains variables z_{i1} and z_{i2}. For each box $X^{(i)}$ to which the vector x belongs, make $z_{i1} =$ "false" if the box has $[0,0]$ on the side associated with variable z_{i1}. Similarly, we make $z_{i2} =$ "false" if the box has $[0,0]$ on the side associated with variable z_{i2}. Because of the way the boxes were build, the Boolean vector we build will make the clause associated with the box corresponding box $X^{(i)}$ false.

For example, if the clause is $z_{i1} \vee z_{i2}$, then the box will have $[0,0]$ for the sides associated with both variable, so they will be both assigned the "false" Boolean value, making the clause false. This means that the Boolean formula built will make at least $n - k$ clauses become false. This formula will satisfy at most $k = n - (n - k)$ clauses.

In the opposite direction, if there is a Boolean vector z which satisfies at most k clauses of the formula F, build a vector $x = (x_1, \ldots, x_n)$ which has value:

- $x_i = 0$ in dimension i if the Boolean variable z_i associated with this dimension is false, and
- $x_i = 1$ otherwise.

One can check that for this arrangement, $x \in X^{(i)}$ if and only if the original Boolean vector z made the corresponding clause C_i false.

Since the Boolean vector z satisfies at most k clauses of the formula F, it makes at least $n - k$ clauses false. This means that the vector x that we have built will belong to all the boxes associated with at least $n - k$ clauses that are false.

The reduction is proven, and so is NP-hardness.

Chapter 5
Towards Better Ways of Extracting Information about Uncertainty from Data

Formulations, results, and methods described in the previous chapters are based on the idealized assumption that we have a good description of the uncertainty of the original data. In practice, often, we do not have this information, we need to extract it from the data. In this final chapter, we describe how this uncertainty information can be extracted from the data.

The results presented in this chapter first appeared in [54, 67, 68].

5.1 Extracting Uncertainty from Data: Traditional Approach and Its Limitations

Need to Fuse Models: Reminder. As we have mentioned in Chapter 4, often, there exist several methods for estimating a certain quantity. For example, in geosciences, to determine density, we can use seismic data [20], we can use gravity measurement, etc. Each of these techniques has its own advantages and limitations: e.g., seismic measurements often lead to a more accurate value of ρ than gravity measurements, but seismic measurements mostly provide information about the areas above the Moho surface. It is desirable to combine ("fuse") the models obtained from different types of measurements into a single model that would combine the advantages of all of these models.

Similar situations are frequent in practice: we are interested in the value of a quantity, and we have reached the limit of the accuracy that can be achieved by using a single available measuring instrument. In this case, to further increase the estimation accuracy, we perform several measurements of the desired quantity x_i – by using the same measuring instrument or different measuring instruments – and combine the results $x_{i1}, x_{i2}, \ldots, x_{im}$ of these measurement into a single more accurate estimate $\widehat{x_i}$; see, e.g., [58, 71].

Traditional Approach Uses Normal Distributions. The need for fusion appears when we have already extracted as much accuracy from each type of measurements as possible.

© Springer International Publishing Switzerland 2015
C. Servin & V. Kreinovich, *Propagation of Interval & Probabilistic Uncertainty*,
Studies in Systems, Decision and Control 15, DOI: 10.1007/978-3-319-12628-9_5

In many practical situations, this means, in particular, that we have found and eliminated the systematic errors (thus, the resulting measurement error has 0 mean), and that we have found and eliminated the major sources of the random error. Since all big error components are eliminated, what is left is the large number of small error components. According to the Central Limit Theorem, the distribution of the sum of a large number of independent small random variables is approximately normal. Thus, it is natural to assume that each measurement error $\Delta x_{ij} \overset{\text{def}}{=} x_{ij} - x_i$ is normally distributed with 0 mean and some variance σ_j^2. Then, the probability density corresponding to x_{ij} is $\dfrac{1}{\sqrt{2\pi} \cdot \sigma_j} \cdot \exp\left(-\dfrac{(x_{ij} - x_i)^2}{2\sigma_j^2}\right)$.

It is also reasonable to assume that measurement errors corresponding to different measurements are independent. Under this assumption, the overall probability density is equal to the product of the corresponding probability distributions

$$L = \prod_{j=1}^{m} \frac{1}{\sqrt{2\pi} \cdot \sigma_j} \cdot \exp\left(-\frac{(x_{ij} - x_i)^2}{2\sigma_j^2}\right). \tag{5.1}$$

According to the Maximum Likelihood Principle, we select the value x_i for which the above probability L is the largest possible. Since $\exp(a) \cdot \exp(b) = \exp(a+b)$, we get

$$L = \prod_{j=1}^{m} \frac{1}{\sqrt{2\pi} \cdot \sigma_j} \cdot \exp\left(-\sum_{j=1}^{m} \frac{(x_{ij} - x_i)^2}{2\sigma_j^2}\right). \tag{5.2}$$

Maximizing L is equivalent to minimizing $-\ln(L)$, i.e., to minimizing the sum $\sum_{j=1}^{m} \dfrac{(x_{ij} - x_i)^2}{\sigma_j^2}$. Differentiating this sum w.r.t. x_i and equating the derivative to 0, we conclude that $\sum_{j=1}^{m} \sigma_j^{-2} \cdot (x_{ij} - x_i) = 0$, so

$$x_i = \frac{\sum_{j=1}^{m} \sigma_j^{-2} \cdot x_{ij}}{\sum_{j=1}^{m} \sigma_j^{-2}}. \tag{5.3}$$

As we have mentioned in Chapter 4, this idea has been successfully applied to geophysics.

Need to Estimate Accuracy of the Corresponding Models. To apply the above formula, we need to know the accuracies σ_j of different models. In this section, we describe the traditional methods of estimating accuracy (see, e.g., [58]) and explain their limitations.

First Method: Calibration. The first method is to *calibrate* the corresponding measuring instrument. Calibration is possible when we have a "standard" measuring instrument which is several times more accurate than the instrument which we are calibrating. We then repeatedly measure the same quantity by using both our measuring instrument and the standard one. Since the standard instrument is much more accurate than the one we testing, the result $x_{i,st}$ of using this instrument is practically equal to the actual value x_i, and thus, the measurement error $\Delta x_{ij} = x_{ij} - x_i$ is well approximated by the difference $\Delta x_{ij} \approx x_{ij} - x_{i,st}$ between the measurement results.

Since all the measurements x_{ij}, $i = 1 \ldots, n$, are performed by the same measuring instrument j, all these measurements have the same standard deviation σ_j. In this case, the likelihood (5.2) take the simplified form

$$L = \frac{1}{(\sqrt{2\pi})^n \cdot \sigma_j^n} \cdot \exp\left(-\sum_{i=1}^{n} \frac{(x_{ij} - x_i)^2}{2\sigma_j^2}\right). \tag{5.4}$$

We need to find the value σ_j for which the likelihood L attains the largest possible value. Maximizing L is equivalent to minimizing $-\ln(L) = \text{const} + n \cdot \ln(\sigma_j) + \sum_{i=1}^{n} \frac{(x_{ij} - x_i)^2}{2\sigma_j^2}$. Differentiating this sum w.r.t. σ_j and equating the derivative to 0, we get the usual estimate

$$\sigma_j^2 = \frac{1}{n} \cdot \sum_{i=1}^{n} (x_{ij} - x_i)^2. \tag{5.5}$$

Since we know approximate values of $x_{ij} - x_i$, we can thus estimate σ_j.

It Is Not Possible to Directly Use Calibration for Estimating Uncertainty of State-of-the Art Measurements. For calibration to work, we need to have a measuring instrument which is several times more accurate than the one that we currently use. The above traditional approach works well for many measuring instruments. However, we cannot apply this approach for calibrating state-of-the-art instrument, because these instruments are the best we have. There are no other instruments which are much more accurate than these ones – and which can therefore serve as standard measuring instruments for our calibration.

Such situations are typical in the applications analyzed by the Cyber-ShARE Center; for example:

- in the environmental sciences, we want to gauge the accuracy with which the Eddy covariance tower measure the Carbon and heat fluxes; see, e.g., [1]
- in the geosciences, we want to gauge how accurately seismic [20], gravity, and other techniques reconstruct the density at different depths and different locations.

Second Method: Using Several Similar Instruments. In some practical situations, when we do have a standard measuring instrument, we can instead compare the results x_{i1} and x_{i2} of using two similar measuring instruments to measure the same quantities x_i. The two instruments are independent and have the same accuracy σ, so the likelihood function has the form

$$L = \frac{1}{(\sqrt{2\pi})^n \cdot \sigma^n} \cdot \exp\left(-\sum_{i=1}^{n} \frac{(x_{i1} - x_i)^2}{2\sigma^2}\right) \cdot \frac{1}{(\sqrt{2\pi})^n \cdot \sigma^n} \cdot \exp\left(-\sum_{i=1}^{n} \frac{(x_{i2} - x_i)^2}{2\sigma^2}\right).$$

In this case, we do not know σ and we do not know the actual values $x_1,\ldots,x_m$; in the spirit of the Maximum Likelihood method, we will select the values of all these parameters for which the likelihood attains the largest possible value. Maximizing L is equivalent to minimizing

$$-\ln(L) = \text{const} + 2n \cdot \ln(\sigma) + \sum_{i=1}^{n} \frac{(x_{i1} - x_i)^2}{2\sigma^2} + \sum_{i=1}^{m} \frac{(x_{i2} - x_i)^2}{2\sigma^2}. \tag{5.6}$$

Minimizing with respect to x_i leads to $x_i = \dfrac{x_{i1} + x_{i2}}{2}$. Substituting these values x_i into the formula (5.6) and minimizing the resulting expression with respect to σ, we get

$$\sigma^2 = \frac{1}{2n} \cdot \sum_{i=1}^{n} (x_{i1} - x_{i2})^2. \tag{5.7}$$

It Is Not Possible to Directly Use This Method Either. In usual measurements, when we estimate the accuracy of measurements performed by a measuring instrument, we can produce two similar measuring instruments and compare their results. In geophysics, we want to estimate the accuracy of a model, e.g., a seismic model, a gravity-based model, etc. In this situation, we do not have two similar applications of the same model, so the second method cannot be directly applied either.

Moreover, Maximum Likelihood Approach Cannot Be Applied to Estimate Model Accuracy. Let us now consider the most general situation: we have several quantities with (unknown) actual values $x_1,\ldots,x_i,\ldots,x_n$, we have several measuring instruments (or geophysical methods) with (unknown) accuracies $\sigma_1,\ldots,\sigma_j,\ldots,\sigma_m$, and we know the results x_{ij} of measuring the i-th quantity by using the j-th measuring instrument. At first glance, a reasonable idea is to find all the unknown quantities – i.e., the actual values x_i and the σ_j – from the Maximum Likelihood method. In this case, the likelihood takes the form

$$L = \prod_{i=1}^{n} \prod_{j=1}^{m} \frac{1}{\sqrt{2\pi} \cdot \sigma_j} \cdot \exp\left(-\frac{(x_{ij} - x_i)^2}{2\sigma_j^2}\right). \tag{5.8}$$

The problem with this approach is that, in contrast to the previous cases, this expression does not attain a finite maximum, it can reach values which are as large as possible. Namely, if we pick some j_0 and take $x_i = x_{ij_0} + \varepsilon$ and $\sigma_{j_0} = \varepsilon$, then we get $\dfrac{(x_{ij_0} - x_i)^2}{2\sigma_{j_0}^2} = \dfrac{1}{2}$, so the corresponding exponential factor is equal to $\exp\left(-\dfrac{1}{2}\right)$; all other factors are also finite (and positive) in the limit $\varepsilon \to 0$ except for the terms $\dfrac{1}{\sqrt{2\pi} \cdot \sigma_{j_0}}$ which tends to infinity.

One can check that if all the values σ_j are positive, then the above likelihood expression attains finite values. Thus, the largest possible – infinite – value is attained when one of the standard deviations σ_{j_0} is equal to 0. In this case, in accordance with the formula (5.3), we get $x_i = x_{ij_0}$. In other words, for this problem, the Maximum Likelihood method leads to a counterintuitive conclusion that one of the measurements was absolutely accurate. This is not physically reasonable, so Maximum Likelihood method cannot be directly used to estimate random errors.

5.2 How to Calibrate State-of-the-Art Measuring Instruments: Case of Normally Distributed Measurement Errors

Analysis of the Problem. We know that $x_{ij} = x_i + \Delta x_{ij}$, where approximation errors $\Delta x_{ij} = x_{ij} - x_i$ are independent normally distributed random variables with 0 mean and (unknown) standard deviations σ_j^2. For every two estimation methods (e.g., measuring instruments) j and k, the difference $x_{ij} - x_{ik}$ between the results of estimating the same quantity x_i by these two methods has the form

$$x_{ij} - x_{ik} = (x_i + \Delta x_{ij}) - (x_i + \Delta x_{ik}) = \Delta x_{ij} - \Delta x_{ik}.$$

Derivation of the Resulting Formula. The difference between two independent normally distributed random variables Δx_{ij} and Δx_{ik} is also normally distributed. The mean of the difference is equal to the difference of the means, i.e., to $0 - 0 = 0$, and the variance of the difference is equal to the sum of the variances, i.e., to $\sigma_j^2 + \sigma_k^2$.

Thus, the difference $x_{ij} - x_{ik} = \Delta x_{ij} - \Delta x_{ik}$ is normally distributed with 0 mean and variance $\sigma_j^2 + \sigma_k^2$. For each j and k, we have n values $x_{1j} - x_{1k}, \ldots, x_{nj} - x_{nk}$ from this distribution. Based on this sample, we can apply the usual formula (5.5) to estimate the standard deviation $\sigma_j^2 + \sigma_k^2$ as $\sigma_j^2 + \sigma_k^2 \approx A_{jk}$, where

$$A_{jk} \stackrel{\text{def}}{=} \frac{1}{n} \cdot \sum_{i=1}^{n} (x_{ij} - x_{ik})^2. \tag{5.9}$$

In particular, for every three different measuring instruments, with unknown accuracies σ_1^2, σ_2^2, and σ_3^2, we get the equations

$$\sigma_1^2 + \sigma_2^2 \approx A_{12}, \quad \sigma_1^2 + \sigma_3^2 \approx A_{13}, \quad \sigma_2^2 + \sigma_3^2 \approx A_{23}. \tag{5.10}$$

By adding all three equalities (5.10) and dividing the result by two, we get

$$\sigma_1^2 + \sigma_2^2 + \sigma_3^2 \approx \frac{A_{12} + A_{13} + A_{23}}{2}. \tag{5.11}$$

Resulting Formulas. Subtracting, from (5.11), each of the equalities (5.10), we conclude that $\sigma_j^2 \approx \widetilde{V}_j$, where

$$\tilde{V}_1 = \frac{A_{12} + A_{13} - A_{23}}{2}; \quad \tilde{V}_2 = \frac{A_{12} + A_{23} - A_{13}}{2}; \quad \tilde{V}_3 = \frac{A_{13} + A_{23} - A_{12}}{2}. \quad (5.12)$$

Comment. In general, when we have M different models, we have $\dfrac{M \cdot (M-1)}{2}$ different equations $\sigma_j^2 + \sigma_k^2 \approx A_{jk}$ to determine N unknowns σ_j^2. When $M > 3$, we have more equations than unknowns, so we can use the Least Squares method to estimate the desired values σ_j^2.

Challenge. The formulas $\sigma_i^2 \approx \tilde{V}_i$ are approximate. If we use an estimate $\tilde{V}_j$ for σ_j^2, we may get physically meaningless negative values for the corresponding variances.

It is therefore necessary to modify the formulas (5.12) so as to avoid negative values.

An Idea of How to Deal with This Challenge. The negativity challenge is caused by the fact that the estimates in (5.12) are approximate. So, to come up with the desired modification, we will first estimate the accuracy of each of the formulas (5.12), i.e., the standard deviation Δ_j for the difference $\Delta V_j \stackrel{\text{def}}{=} \tilde{V}_j - \sigma_j^2$.

For large n, the difference ΔV_j between the actual value of σ_j^2 and its statistical estimate is asymptotically normally distributed, with asymptotically 0 mean; see, e.g., [71]. In the next section, we will estimate the standard deviation Δ_j for this difference. Thus, we can conclude that the actual value $\sigma_j^2 = \tilde{V}_j - \Delta V_j$ is normally distributed with mean V_j and standard deviation Δ_j. We also know that $\sigma_j^2 \geq 0$. As an estimate for σ_j^2, it is therefore reasonable to use a conditional expected value $E\left(\tilde{V}_j - \Delta V_j \,\middle|\, \tilde{V}_j - \Delta V_j \geq 0\right)$. This new estimate is an expected value of a non-negative number and thus, cannot be negative. Let us show how to compute this new estimate.

Estimating Accuracies Δ_j of the Estimates $\overline{V}_j$ for σ_j^2. Let us begin by estimating the accuracy Δ_j of $\tilde{V}_j$, i.e., the expected value $\Delta_j^2 = E\left[\left(\tilde{V}_j - \sigma_j^2\right)^2\right]$. According to (5.12), $\tilde{V}_j$ is computed based on the values

$$A_{jk} = \frac{1}{n} \cdot \sum_{i=1}^{n}(x_{ij} - x_{ik})^2 = \frac{1}{n} \cdot \sum_{i=1}^{n}(\Delta x_{ij} - \Delta x_{ik})^2.$$

To simplify notations, let us denote $a_i \stackrel{\text{def}}{=} \Delta x_{ij}$, $b_i \stackrel{\text{def}}{=} \Delta x_{ik}$, and $c_i \stackrel{\text{def}}{=} \Delta x_{i\ell}$; then, we conclude that

$$\tilde{V}_j = \frac{1}{2} \cdot \left[\frac{1}{n} \cdot \sum_{i=1}^{n}(a_i - b_i)^2 + \frac{1}{n} \cdot \sum_{i=1}^{n}(a_i - c_i)^2 - \frac{1}{n} \cdot \sum_{i=1}^{n}(b_i - c_i)^2 \right],$$

i.e.,

$$\tilde{V}_j = \frac{1}{2n} \cdot \sum_{i=1}^{n}\left[(a_i - b_i)^2 + (a_i - c_i)^2 - (b_i - c_i)^2\right]. \quad (5.13)$$

Opening parentheses inside the sum, we get

$$(a_i - b_i)^2 + (a_i - c_i)^2 - (b_i - c_i)^2 = a_i^2 - 2a_i \cdot b_i + b_i^2 + a_i^2 - 2a_i \cdot c_i + c_i^2 - b_i^2 + 2b_i \cdot c_i - c_i^2.$$

Thus, the formula (5.13) takes the form

$$\tilde{V}_j = \frac{1}{n} \cdot \sum_{i=1}^{n} (a_i^2 - a_i \cdot b_i - a_i \cdot c_i + b_i \cdot c_i).$$

Therefore,

$$\Delta_j^2 = E\left[\left(\tilde{V}_j - \sigma_j^2\right)\right] = E\left[\left(\tilde{V}_j\right)^2 - 2\tilde{V}_j \cdot \sigma_j^2 + \sigma_j^4\right] = E_1 - 2\sigma_1^2 \cdot E_2 + \sigma_1^4, \quad (5.14)$$

where

$$E_1 \overset{\text{def}}{=} E\left[\left(\tilde{V}_j\right)^2\right] = E\left[\left(\frac{1}{n} \cdot \sum_{i=1}^{n} (a_i^2 - a_i \cdot b_i - a_i \cdot c_i + b_i \cdot c_i)\right)^2\right], \quad (5.15)$$

$$E_2 \overset{\text{def}}{=} E\left[\tilde{V}_j\right] = E\left[\frac{1}{n} \cdot \sum_{i=1}^{n} (a_i^2 - a_i \cdot b_i - a_i \cdot c_i + b_i \cdot c_i)\right].$$

The expected value E_2 is equal to linear combination of the expected values of the expressions a_i^2, $a_i \cdot b_i$, $a_i \cdot c_i$, and $b_i \cdot c_i$:

$$E_2 = \frac{1}{n} \cdot \sum_{i=1}^{n} \left(E[a_i^2] - E[a_i \cdot b_i] - E[a_i \cdot c_i] + E[b_i \cdot c_i]\right). \quad (5.16)$$

All variables a_i, b_i, and c_i are independent and normally distributed with 0 mean and the corresponding variances $V_j = \sigma_j^2$. Due to independence, $E[a_i \cdot b_i] = E[a_i] \cdot E[b_i] = 0 \cdot 0 = 0$; similarly $E[a_i \cdot c_i] = E[b_i \cdot c_i] = 0$, and the only non-zero term is $E[a_i^2] = \sigma_j^2$. Thus, in the sum in E_2, only n terms $a_1^2, \ldots, a_n^2$ lead to non-zero expected value σ_j^2, hence $E_2 = \frac{1}{n} \cdot n \cdot \sigma_j^2 = \sigma_j^2$.

Let us now compute E_1. In general, the square of a sum can be represented as $\left(\sum_i z_i\right)^2 = \sum_i z_i^2 + \sum_{i \neq i'} z_i \cdot z_{i'}$. In our case, $z_i = a_i^2 - a_i \cdot b_i - a_i \cdot c_i + b_i \cdot c_i$. Thus, the expected value E_2 can be presented as

$$E_1 = \frac{1}{n^2} \cdot \sum_{i=1}^{n} E[z_i^2] + \frac{1}{n^2} \cdot \sum_{i \neq i'} E[z_i \cdot z_{i'}]. \quad (5.17)$$

Here, the expression $z_i^2 = \left(a_i^2 - a_i \cdot b_i - a_i \cdot c_i + b_i \cdot c_i\right)^2$ takes the form

$$z_i^2 = a_i^4 + a_i^2 \cdot b_i^2 + a_i^2 \cdot c_i^2 + b_i^2 \cdot c_i^2 + \text{ terms which are odd in } a_i, b_i, \text{ or } c_i.$$

Due to independence and the fact that all normally distributed variables a_i, b_i, and c_i have 0 mean and thus, 0 odd moments, the expected values of odd terms like $a_i^3 \cdot b_i$ is zero: e.g., $E[a_i^3 \cdot b_i] = E[a_i^3] \cdot E[b_i] = 0$. Thus,

$$E[z_i^2] = E[a_i^4] + E[a_i^2 \cdot b_i^2] + E[a_i^2 \cdot c_i^2] + E[b_i^2 \cdot c_i^2].$$

For the normal distribution, $E[a_i^4] = 3\sigma_j^4$; due to independence, $E[a_i^2 \cdot b_i^2] = E[a_i^2] \cdot E[b_i^2] = \sigma_j^2 \cdot \sigma_k^2$. Thus,

$$E[z_i^2] = 3\sigma_j^4 + \sigma_j^2 \cdot \sigma_k^2 + \sigma_j^2 \cdot \sigma_\ell^2 + \sigma_k^2 \cdot \sigma_\ell^2,$$

and

$$\frac{1}{n^2} \cdot \sum_{i=1}^{n} E[z_i^2] = \frac{1}{n} \cdot (3\sigma_j^4 + \sigma_j^2 \cdot \sigma_k^2 + \sigma_j^2 \cdot \sigma_\ell^2 + \sigma_k^2 \cdot \sigma_\ell^2). \tag{5.18}$$

For $z_i \cdot z_{i'}$ with $i \neq i'$, we similarly have

$$z_i \cdot z_{i'} = (a_i^2 - a_i \cdot b_i - a_i \cdot c_i + b_i \cdot c_i) \cdot (a_{i'}^2 - a_{i'} \cdot b_{i'} - a_{i'} \cdot c_{i'} + b_{i'} \cdot c_{i'}) = a_i^2 \cdot a_{i'}^2 +$$

$$\text{odd terms with 0 mean.}$$

Thus, $E[z_i \cdot z_{i'}] = E[a_i^2 \cdot a_{i'}^2] = E[a_i^2] \cdot E[a_{i'}^2] = \sigma_j^2 \cdot \sigma_j^2 = \sigma_j^4$ and so, after adding over all $n^2 - n$ pairs (i, i') with $i \neq i'$, we get

$$\frac{1}{n^2} \cdot \sum_{i \neq i'}^{n} E[z_i \cdot z_{i'}] = \frac{n^2 - n}{n^2} \cdot \sigma_j^4 = \left(1 - \frac{1}{n}\right) \cdot \sigma_j^4. \tag{5.19}$$

Substituting the expressions (5.18) and (5.19) into the formula (5.17), we conclude that

$$E_1 = \frac{1}{n} \cdot (3\sigma_j^4 + \sigma_j^2 \cdot \sigma_k^2 + \sigma_j^2 \cdot \sigma_\ell^2 + \sigma_k^2 \cdot \sigma_\ell^2) + \left(1 - \frac{1}{n}\right) \cdot \sigma_j^4.$$

Substituting this expression for E_1 and the formula $E_2 = \sigma_j^2$ into the formula (5.14), we get

$$\Delta_j^2 = \frac{1}{n} \cdot (3\sigma_j^4 + \sigma_j^2 \cdot \sigma_k^2 + \sigma_j^2 \cdot \sigma_\ell^2 + \sigma_k^2 \cdot \sigma_\ell^2) + \left(1 - \frac{1}{n}\right) \cdot \sigma_j^4 - 2\sigma_j^4 + \sigma_j^4,$$

i.e.,

$$\Delta_j^2 = \frac{1}{n} \cdot (2\sigma_j^4 + \sigma_j^2 \cdot \sigma_k^2 + \sigma_j^2 \cdot \sigma_\ell^2 + \sigma_k^2 \cdot \sigma_\ell^2). \tag{5.20}$$

We do not know the exact values σ_j^2, but we do no know the estimates $\widetilde{V}_j$ for these values; thus, we can estimate Δ_j as follows:

$$\Delta_j^2 \approx \frac{1}{n} \cdot \left(\left(\widetilde{V}_j\right)^2 + \widetilde{V}_j \cdot \widetilde{V}_k + \widetilde{V}_j \cdot \widetilde{V}_\ell + \widetilde{V}_k \cdot \widetilde{V}_\ell \right). \tag{5.21}$$

From Estimating Δ_j to a Non-negative Estimate for σ_j^2. So far, we have an estimate $\tilde{V}_j$ for σ_j^2 (as defined by the formula (5.12)), we know that the difference $\Delta V_j = \tilde{V}_j - \sigma_j^2$ is normally distributed with 0 mean, and we know the standard deviation Δ_j of this difference. Since, as we mentioned in the previous section, the original estimate $\tilde{V}_j$ may be negative, it is desirable to use a new estimate $E\left(\tilde{V}_j - \Delta V_j \,\middle|\, \tilde{V}_j - \Delta V_j \geq 0\right)$.

The Gaussian variable ΔV_j has 0 mean and standard deviation Δ_j; thus, it can be represented as $t \cdot \Delta_j$, where t is a Gaussian random variable with 0 and standard deviation 1. In terms of the new variable t, the non-negativity condition $\tilde{V}_j - \Delta V_j \geq 0$ takes the form $\tilde{V}_j - \Delta_j \cdot t \geq 0$, i.e., $t \leq \delta_j \overset{\text{def}}{=} \dfrac{\tilde{V}_j}{\Delta_j}$. Thus, the desired conditional mean is equal to

$$E\left(\tilde{V}_j - \Delta_j \cdot t \,\middle|\, t \leq \delta_j\right) = E\left(\tilde{V}_j \,\middle|\, t \leq \delta_j\right) - \Delta_j \cdot E\left(t \,\middle|\, t \leq \delta_j\right) =$$

$$\tilde{V}_j - \Delta_j \cdot E\left(t \,\middle|\, t \leq \delta_j\right). \tag{5.22}$$

So, to compute the desired estimate, it is sufficient to be able to compute the value $E\left(t \,\middle|\, t \leq \delta_j\right)$ for the standard Gaussian variable t, with the probability density function $\rho(t) = \dfrac{1}{\sqrt{2\pi}} \cdot \exp\left(-\dfrac{t^2}{2}\right)$. By definition, this conditional mean is equal to the ratio $E\left(t \,\middle|\, t \leq \delta_j\right) = \dfrac{N_j}{D_j}$, where

$$N_j = \int_{-\infty}^{\delta_j} t \cdot \rho(t)\, dt; \quad D_j = \int_{-\infty}^{\delta_j} \rho(t)\, dt. \tag{5.23}$$

The denominator D_j is equal to $\Phi(\delta_j) \overset{\text{def}}{=} \text{Prob}(t \leq \delta_j)$. The numerator N_j of this formula is equal to

$$N_j = \int_{-\infty}^{\delta_j} t \cdot \frac{1}{\sqrt{2\pi}} \cdot \exp\left(-\frac{t^2}{2}\right) dt. \tag{5.24}$$

By introducing a new variable $s = \dfrac{t^2}{2}$ for which $ds = t \cdot dt$, we reduce (5.24) to

$$N_j = \frac{1}{\sqrt{2\pi}} \cdot \int_{\infty}^{\delta_j^2/2} \exp(-s)\, ds.$$

This integral can be explicitly computed, so we get

$$N_j = -\frac{1}{\sqrt{2\pi}} \cdot \exp\left(-\frac{\delta_j^2}{2}\right)$$

and thus,

$$E\left(t \mid t \le \delta_j\right) = -\frac{1}{\sqrt{2\pi}} \cdot \frac{\exp\left(-\dfrac{\delta_j^2}{2}\right)}{\Phi(\delta_j)}.$$

So,

$$E\left(\tilde{V}_j - \Delta_j \cdot t \mid t \le \delta_j\right) = \tilde{V}_j - \Delta_j \cdot E\left(t \mid t \le \delta_j\right) = \tilde{V}_j + \frac{\Delta_j}{\sqrt{2\pi}} \cdot \frac{\exp\left(-\dfrac{\delta_j^2}{2}\right)}{\Phi(\delta_j)}.$$

Resulting Algorithm. Let us assume that for each value x_i ($i = 1, \ldots, n$), we have three estimates x_{i1}, x_{i2}, and x_{i3} corresponding to three different models. Our objective is to estimate the accuracies σ_j^2 of these three models.

First, for each $j \ne k$, we compute $A_{jk} = \dfrac{1}{n} \cdot \sum_{i=1}^{n} (x_{ij} - x_{ik})^2$. Then, we compute

$$\tilde{V}_1 = \frac{A_{12} + A_{13} - A_{23}}{2}; \quad \tilde{V}_2 = \frac{A_{12} + A_{23} - A_{13}}{2}; \quad \tilde{V}_3 = \frac{A_{13} + A_{23} - A_{12}}{2}.$$

After that, for each j, we compute

$$\Delta_j^2 = \frac{1}{n} \cdot \left(\left(\tilde{V}_j\right)^2 + \tilde{V}_j \cdot \tilde{V}_k + \tilde{V}_j \cdot \tilde{V}_\ell + \tilde{V}_k \cdot \tilde{V}_\ell\right).$$

Once we compute the preliminary estimates $\tilde{V}_j$ and their accuracies Δ_j, we then compute the auxiliary ratios $\delta_j = \dfrac{\tilde{V}_j}{\Delta_j}$ and return, as an estimate $\widetilde{\sigma_j^2}$ for σ_j^2, the value

$$\widetilde{\sigma_j^2} = \tilde{V}_j + \frac{\Delta_j}{\sqrt{2\pi}} \cdot \frac{\exp\left(-\dfrac{\delta_j^2}{2}\right)}{\Phi(\delta_j)}.$$

5.3 How to Calibrate State-of-the-Art Measuring Instruments: General Case

Need to go beyond Normal Distributions, and Resulting Problem. In practice, the distribution of measurement errors is often different from normal; this is the case, e.g., in measuring fluxes [1]. In such cases, we can still use the same techniques to find the standard deviation of the measurement error. However, in general, it is not enough to know the standard deviation to uniquely determine the distribution: e.g., we may have (and we sometimes do have) an asymmetric distribution, for which the skewness is different from 0 (i.e., equivalently, the expected value of $(\Delta x)^3$ is different from 0).

It is known that in this case, in contrast to the case of the normal distribution, we cannot uniquely reconstruct the distribution of Δx from the known distribution of the difference $\Delta x^{(1)} - \Delta x^{(2)}$ (see, e.g., [1]). Indeed, if we have an asymmetric distribution for Δx, i.e., a distribution which is not invariant under the transformation $\Delta x \to -\Delta x$, this means that the distribution for $\Delta y \stackrel{\text{def}}{=} -\Delta x$ is different from the distribution for Δx. However, since

$$\Delta y^{(1)} - \Delta y^{(2)} = \Delta x^{(2)} - \Delta x^{(1)},$$

the y-difference is also equal to the difference between two independent variables with the distribution Δx and thus, distribution for the difference $\Delta y^{(1)} - \Delta y^{(2)}$ is exactly the same as for the difference $\Delta x^{(1)} - \Delta x^{(2)}$. In other words, if we know the distribution for the difference $\Delta x^{(1)} - \Delta x^{(2)}$, we cannot uniquely reconstruct the distribution for Δx, because, in addition to the original distribution for Δx, all the observations are also consistent with the distribution for $\Delta y = -\Delta x$.

This known non-uniqueness naturally leads to the following questions:

- first, a theoretical question: since we cannot uniquely reconstruct the distribution for Δx, what information about this distribution can we reconstruct?
- second, a practical question: for those characteristics of Δx which can be theoretically reconstructed, we need to design computationally efficient algorithms for reconstructing these characteristics.

What We Are Planning to Do. In this chapter, we address these two questions.

5.4 Techniques for Solving the Problem

To answer the above questions, we use the Fourier analysis technique – the techniques that we have already used in Chapter 2.

What we want to find is the probability density $\rho(z)$ describing the distribution of the measurement error $z \stackrel{\text{def}}{=} \Delta x$. In order to find the unknown probability density, we will first find its Fourier transform

$$F(\omega) = \int \rho(z) \cdot e^{i \cdot \omega \cdot z} \, dz.$$

By definition, this Fourier transform is equal to the mathematical expectation of the function $e^{i \cdot \omega \cdot z}$:

$$F(\omega) = E\left[e^{i \cdot \omega \cdot z} \right].$$

Such a mathematical expectation is also known as a *characteristic function* of the random variable z.

Based on the observed values of the difference $z^{(1)} - z^{(2)}$, we can estimate the characteristic function $D(\omega)$ of this difference:

$$D(\omega) = E\left[e^{i \cdot \omega \cdot (z^{(1)} - z^{(2)})}\right].$$

Here,

$$e^{i \cdot \omega \cdot (z^{(1)} - z^{(2)})} = e^{(i \cdot \omega \cdot z^{(1)}) + (-i \cdot \omega \cdot z^{(2)})} = e^{i \cdot \omega \cdot z^{(1)}} \cdot e^{-i \cdot \omega \cdot z^{(2)}}.$$

As we have mentioned in Chapter 1, measurement errors $z^{(1)}$ and $z^{(2)}$ corresponding to two measuring instruments are usually assumed to be independent. Thus, the variables $e^{i \cdot \omega \cdot z^{(1)}}$ and $e^{-i \cdot \omega \cdot z^{(2)}}$ are also independent. It is known that the expected value of the product of two independent variables is equal to the product of their expected values, thus,

$$D(\omega) = E\left[e^{i \cdot \omega \cdot z^{(1)}}\right] \cdot E\left[e^{-i \cdot \omega \cdot z^{(2)}}\right],$$

i.e.,

$$D(\omega) = F(\omega) \cdot F(-\omega).$$

Here,

$$F(-\omega) = E\left[e^{-i \cdot \omega \cdot z}\right] = E\left[\left(e^{i \cdot \omega \cdot z}\right)^*\right],$$

where t^* means complex conjugation, i.e., an operation that transforms $t = a + b \cdot i$ into $t^* = a - b \cdot i$. Thus, $F(-\omega) = F^*(\omega)$, and the above formula takes the form

$$D(\omega) = F(\omega) \cdot F^*(\omega) = |F(\omega)|^2.$$

In other words, the fact that we know $D(\omega)$ means that we know the absolute value (modulus) of the complex-valued function $F(\omega)$.

In these terms, the problems becomes: how can we reconstruct the complex-valued function $F(\omega)$ if we only know its absolute value?

5.5   Is It Possible to Estimate Accuracy?

How to Use These Techniques to solve the Theoretical Question. First, let us address the theoretical question: since, in general, we cannot reconstruct $\rho(z)$ (or, equivalently, $F(\omega)$) uniquely, what information about $\rho(z)$ (and, correspondingly, about $F(\omega)$) can we reconstruct?

To solve this theoretical question, let us take into account the practical features of this problem. First, it needs to be mentioned that, from the practical viewpoint, we need to take into account that the situation in, e.g., Eddy covariance tower measurements is more complex that we described, because the tower does not measure *one* single quantity, it simultaneously measuring *several* quantities: carbon flux, heat flux, etc. Since these different measurements are based on data from the same sensors, it is reasonable to expect that the resulting measurement errors are correlated.

Thus, to fully describe the measurement uncertainty, it is not enough to describe the distribution of each 1-D measurement error, we need to describe a joint distribution of all the measurement errors $z = (z_1, z_2, \ldots)$. In this multi-D case, we can use the multi-D Fourier transforms and characteristic functions, where for $\omega = (\omega_1, \omega_2, \ldots)$, we define

$$F(\omega) = E\left[e^{i \cdot \omega \cdot z}\right],$$

with

$$\omega \cdot z \stackrel{\text{def}}{=} \omega_1 \cdot z_1 + \omega_2 \cdot z_2 + \ldots$$

Second, we need to take into account that while theoretically, we can consider all possible values of the difference $z^{(1)} - z^{(2)}$, in practice, we can only get values which are proportional to the smallest measuring unit h. For example, if we measure distance and the smallest distance we can measure is centimeters, then the measuring instrument can only return values 0 cm, 1 cm, 2 cm, etc. In other words, in reality, the value z can only take discrete values. If we take the smallest value of z as the new starting point (i.e., as 0), then the possible values of z take the form $z = 0$, $z = h$, $z = 2h, \ldots$, until we reach the upper bound $z = N \cdot h$ for some integer N. For these values, in the 1-D case, the Fourier transform takes the form

$$F(\omega) = E\left[e^{i \cdot \omega \cdot z}\right] = \sum_{k=0}^{N} p_k \cdot e^{i \cdot \omega \cdot k \cdot h},$$

where p_k is the probability of the value $z = k \cdot h$. This formula can be equivalently rewritten as

$$F(\omega) = \sum_{k=0}^{N} p_k \cdot v^k,$$

where $v \stackrel{\text{def}}{=} e^{i \cdot \omega \cdot h}$. Similarly, in the multi-D case, we have $z = (k_1 \cdot h_1, k_2 \cdot h_2, \ldots)$, and thus,

$$e^{i \cdot \omega \cdot k \cdot h} = e^{i \cdot \omega \cdot (k_1 \cdot h_1 + k_2 \cdot h_2 + \ldots)} = e^{i \cdot \omega_1 \cdot k_1 \cdot h_1} \cdot e^{i \cdot \omega \cdot k_2 \cdot h_2} \cdot \ldots,$$

so we have

$$F(\omega) = \sum_{k_1=0}^{N_1} \sum_{k_2=0}^{N_2} \ldots p_k \cdot v_1^{k_1} \cdot v_2^{k_2} \cdot \ldots,$$

where $v_k \stackrel{\text{def}}{=} e^{i \cdot \omega_k \cdot h_k}$. In other words, we have a polynomial of the variables $v_1, v_2, \ldots$:

$$P(v_1, v_2, \ldots) = \sum_{k_1=0}^{N_1} \sum_{k_2=0}^{N_2} \ldots p_k \cdot v_1^{k_1} \cdot v_2^{k_2} \cdot \ldots$$

Different values of ω correspond to different values of $v = (v_1, v_2, \ldots)$. Thus, the fact that we know the values of $|F(\omega)|^2$ for different ω is equivalent to knowing the values of $|P(v)|^2$ for all possible values $v = (v_1, v_2, \ldots)$.

In these terms, the theoretical question takes the following form: we know the values $D(v) = |P(v)|^2 = P(v) \cdot P^*(v)$ for some polynomial $P(v)$, we need to reconstruct this polynomial. In the 1-D case, each complex-valued polynomial of degree N has, in general, N complex roots $v^{(1)}$, $v^{(2)}$, etc., and can, therefore, be represented as

$$|P(v)|^2 = \text{const} \cdot (v - v^{(1)}) \cdot (v - v^{(2)}) \cdot \ldots$$

In this case, there are many factors, so there are many ways to represent it as a product – which explains the above-described non-uniqueness of representing $D(v)$ as the product of two polynomials $P(v)$ and $P^*(v)$

Interestingly, in contrast to the 1-D case, in which each polynomial can be represented as a product of polynomials of 1st order, in the multi-D case, a generic polynomial *cannot* be represented as a product of polynomials of smaller degrees. This fact can be easily illustrated on the example of polynomials of two variables. To describe a general polynomial of two variables $\sum_{k=0}^{n} \sum_{l=1}^{n} c_{kl} \cdot v_1^k \cdot v_2^l$ in which each of the variables has a degree $\leq n$, we need to describe all possible coefficients c_{kl}. Each of the indices k and l can take $n + 1$ possible values $0, 1, \ldots, n$, so overall, we need to describe $(n + 1)^2$ coefficients.

When two polynomials multiply, the degrees add: $v^m \cdot v^{m'} = v^{m+m'}$. Thus, if we represent $P(v)$ as a product of two polynomials, one of them must have a degree $m < n$, and the other one degree $n - m$. In general:

- we need $(m + 1)^2$ coefficients to describe a polynomial of degree m and
- we need $(n - m + 1)^2$ coefficients to describe a polynomial of degree $n - m$;
- so, to describe arbitrary products of such polynomials, we need

$$(m + 1)^2 + (n - m + 1)^2$$

coefficients.

To be more precise, in such a product, we can always multiply one of the polynomials by a constant and divide another one by the same constant, without changing the product. Thus, we can always assume that, e.g., in the first polynomial, the free term c_{00} is equal to 1. As a result, we need one fewer coefficient to describe a general product: $(m + 1)^2 + (n - m + 1)^2 - 1$.

To be able to represent a generic polynomial $P(v)$ of degree n as such a product

$$P(v) = P_m(v) \cdot P_{n-m}(v),$$

we need to make sure that the coefficients at all all $(n + 1)^2$ possible degrees $v_1^k \cdot v_2^l$ are the same on both sides of this equation. This requirement leads to $(n + 1)^2$ equations with $(m + 1)^2 + (n - m + 1)^2 - 1$ unknowns.

In general, a system of equations is solvable if the number of equations does not exceed the number of unknowns. Thus, we must have

$$(n + 1)^2 \leq (m + 1)^2 + (n - m + 1)^2 - 1.$$

Opening parentheses, we get

$$n^2 + 2n + 1 \leq m^2 + 2m + 1 + (n-m)^2 + 2 \cdot (n-m) + 1 - 1.$$

The constant terms in both sides cancel each other, as well as the terms $2n$ in the left-hand side and $2m + 2 \cdot (n-m) = 2n$ in the right-hand side, so we get an equivalent inequality

$$n^2 \leq m^2 + (n-m)^2.$$

Opening parentheses, we get

$$n^2 \leq m^2 + n^2 - 2 \cdot n \cdot m + m^2.$$

Canceling n^2 in both sides, we get

$$0 \leq 2m^2 - 2 \cdot n \cdot m.$$

Dividing both sides by $2m$, we get an equivalent inequality $0 \leq m - n$, which clearly contradicts to our assumption that $m < n$.

Let us go back to our problem. We know the product $D(v) = P(v) \cdot P^*(v)$, and we want to reconstruct the polynomial $P(v)$. We know that this problem is not uniquely solvable, i.e., that there exist other polynomials $Q(v) \neq P(v)$ for which $D(v) = P(v) \cdot P^*(v) = Q(v) \cdot Q^*(v)$. Since, in general, a polynomial $P(v)$ of several variables cannot be represented as a product – i.e., is "prime" in terms of factorization the same way prime numbers are – the fact that the two products coincide means that $Q(v)$ must be equal to one of the two prime factors in the decomposition $D(v) = P(v) \cdot P^*(v)$. Since we know that $Q(v)$ is different from $P(v)$, we thus conclude that $Q(v) = P^*(v)$.

By going back to the definitions, one can see that for the distribution $\rho'(x) = \rho(-x)$, the corresponding polynomial has exactly the form $Q(v) = P^*(v)$. Thus, in general, this is the *only* non-uniqueness that we have: each distribution which is consistent with the observation of differences coincides either with the original distribution $\rho(x)$ or with the distribution $\rho'(x) = \rho(-x)$.

In other words, we arrive at the following result.

Answer to the Theoretical Question. We have proven that, in general, each distribution which is consistent with the observation of differences $\Delta x^{(1)} - \Delta x^{(2)}$ coincides either with the original distribution $\rho(x)$ or with the distribution $\rho'(x) \stackrel{\text{def}}{=} \rho(-x)$.

5.6 Practical Question: How to Gauge the Accuracy

How to Use Fourier Techniques to Solve the Practical Question: Idea. We want to find a probability distribution $\rho(z)$ which is consistent with the observed characteristic function $D(\omega)$ for the difference. In precise terms, we want to find a function $\rho(z)$ which satisfies the following two conditions:

- the first condition is that $\rho(z) \geq 0$ for all z, and
- the second condition is $|F(\omega)|^2 = D(\omega)$, where $F(\omega)$ denotes the Fourier transform of the function $\rho(x)$.

One way to find the unknown function that satisfies two conditions is to use the method of successive projections. In this method, we start with an arbitrary function $\rho^{(0)}(z)$. On the k-th iteration, we start with the result $\rho^{(k-1)}(z)$ of the previous iteration, and we do the following:

- first, we project this function $\rho^{(k-1)}(z)$ onto the set of all functions which satisfy the first condition; to be more precise, among all the functions which satisfy the first condition, we find the function $\rho'(x)$ which is the closest to $\rho^{(k-1)}(z)$;
- then, we project the function $\rho'(z)$ onto the set of all functions which satisfy the second condition; to be more precise, among all the functions which satisfy the second condition, we find the function $\rho^{(k)}(x)$ which is the closest to $\rho'(z)$.

We continue this process until it converges.

As the distance between the two functions $f(z)$ and $g(z)$ – describing how close they are – it is reasonable to take the natural analog of the Euclidean distance:

$$d(f,g) \stackrel{\text{def}}{=} \sqrt{\int (f(z) - g(z))^2 \, dz}.$$

One can check that for this distance function:

- the closest function in the first part of the iteration is the function $\rho'(z) = \max(0, \rho^{(k-1)}(z))$, and
- on the second part, the function whose Fourier transform is equal to

$$F^{(k)}(\omega) = \frac{\sqrt{|D(\omega)|}}{|F'(\omega)|} \cdot F'(\omega).$$

Thus, we arrive at the following algorithm.

How to Use Fourier Techniques to Solve the Practical Question: Algorithm.
We start with an arbitrary function $\rho^{(0)}(z)$. On the k-th iteration, we start with the function $\rho^{(k-1)}(z)$ obtained on the previous iteration, and we do the following:

- first, we compute $\rho'(z) = \max(0, \rho^{(k-1)}(z))$;
- then, we apply Fourier transform to $\rho'(z)$ and get $F'(z)$;
- after that, we compute

$$F^{(k)}(\omega) = \frac{\sqrt{|D(\omega)|}}{|F'(\omega)|} \cdot F'(\omega);$$

- finally, as the next approximation $\rho^{(k)}(z)$, we take the result of applying the inverse Fourier transform to $F^{(k)}(\omega)$.

We continue this process until it converges.

5.7 When Can We Reconstruct 1-D Distribution: Numerical Examples

General Idea. In the previous sections, we have shown that a 2-D distribution can be (almost) uniquely reconstructed by comparing the results $x^{(1)}$ and $x^{(2)}$ of measuring the same (unknown) quantity by using two similar measuring instruments. We also described an algorithm for such a reconstruction.

We have also mentioned that, in contrast to the 2-D case, a general 1-D distribution cannot be uniquely reconstructed from the difference $x^{(1)} - x^{(2)}$. This *general* impossibility does not mean, of course, that *some* distributions cannot be thus reconstructed by the same algorithm. To find out whether such a 1-D reconstruction is indeed possible, we simulated 1-D distributions of different complexity, simulated the difference $x^{(1)} - x^{(2)}$, and applied the above algorithm to (try to) reconstruct the original probability density function. The results of these experiments are given in this section.

Simplest Case: A 1-Point Distribution. We started with the simplest case, when the random variable is located at a single point with probability 1. In this case, the above algorithm leads to the perfect reconstruction.

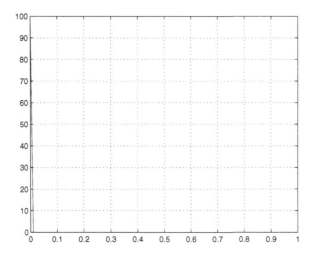

Fig. 5.1 Reconstruction of a 1-point distribution

2-Point Distributions. We then considered 2-point distributions, in which the random variable takes the value 0 with probability p and the value $x > 0$ with probability $1 - p$. We considered values $p = 0.1$, $p = 0.2$, ..., $p = 0.9$. For each of these probability values, we considered $x = 0.1$, $x = 0.2$, ..., $x = 0.9$. In all 81 cases, we also got a perfect reconstruction. To illustrate these results, let us consider the examples corresponding to $p = 0.1$ and to $x = 0.1$.

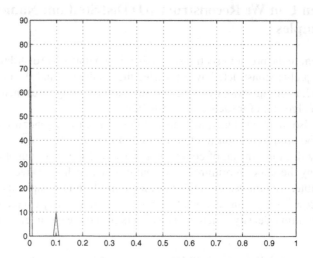

Fig. 5.2 Reconstruction of a 2-point distribution, $p = 0.1$, $x = 0.1$

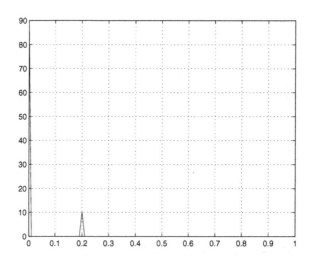

Fig. 5.3 Reconstruction of a 2-point distribution, $p = 0.1$, $x = 0.2$

3-Point Distributions. For 3-point distributions, i.e., distributions in which we have:

- $x = 0$ with probability p_0,
- $x = x_1$ with probability p_1, and
- $x = x_2$ with the remaining probability $1 - p_0 - p_1$,

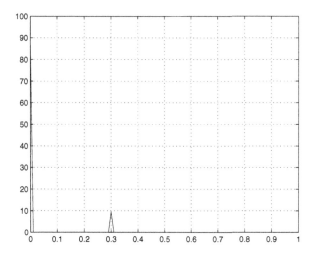

Fig. 5.4 Reconstruction of a 2-point distribution, $p = 0.1, x = 0.3$

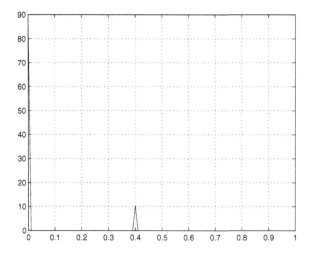

Fig. 5.5 Reconstruction of a 2-point distribution, $p = 0.1, x = 0.4$

our algorithm does not always reconstruct the original distribution for Δx from the observed sample of the differences $\Delta x^{(1)} - \Delta x^{(2)}$. This fact is in good accordance with the fact that in general, our algorithm is only guaranteed to work in the 2-D case.

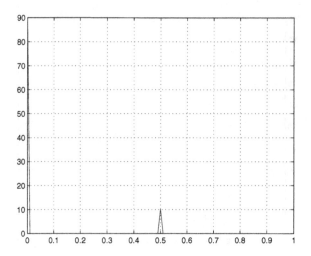

Fig. 5.6 Reconstruction of a 2-point distribution, $p = 0.1$, $x = 0.5$

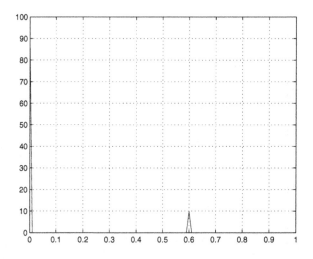

Fig. 5.7 Reconstruction of a 2-point distribution, $p = 0.1$, $x = 0.6$

Here:

- in some situations, we have a good reconstruction;
- in other situations, the reconstructed distribution even has additional peaks.

Below, we give examples of both types of situations.

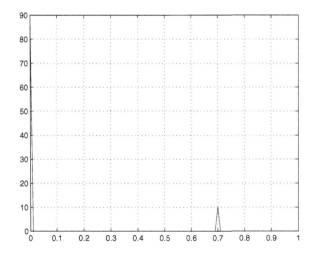

Fig. 5.8 Reconstruction of a 2-point distribution, $p = 0.1$, $x = 0.7$

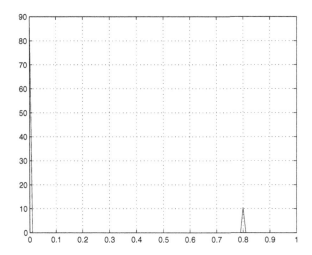

Fig. 5.9 Reconstruction of a 2-point distribution, $p = 0.1$, $x = 0.8$

Continuous Asymmetric Distributions. As an example of a continuous asymmetric distribution, we took a distribution with an asymmetric triangular probability density function $\rho(x) = 2(1 - x)$ for $x \in [0, 1]$ (with $\rho(x) = 0$ outside the interval $[0, 1]$).

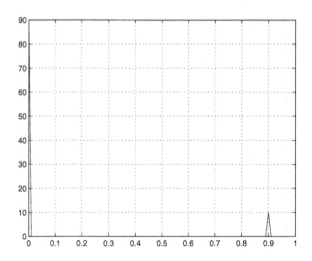

Fig. 5.10 Reconstruction of a 2-point distribution, $p = 0.1, x = 0.9$

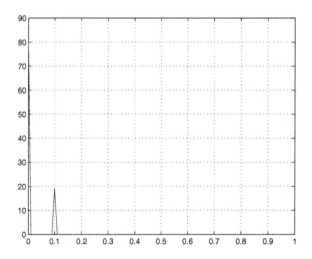

Fig. 5.11 Reconstruction of a 2-point distribution, $p = 0.2, x = 0.1$

Since the reconstruction is based on the randomly generated sample, not surprisingly, the reconstructed distribution shows random deviations from the original one; see Fig. 5.17, on which we also plotted the original distribution.

To decrease these deviations, instead of the reconstructed values $\rho(x)$, we take the average of all the reconstructed values $\rho(x')$ for all values $x' \in [x - \varepsilon, x + \varepsilon]$ for some small $\varepsilon > 0$ (we took $\varepsilon = 0.05$).

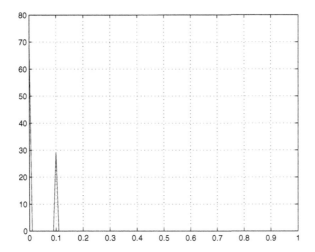

Fig. 5.12 Reconstruction of a 2-point distribution, $p = 0.3, x = 0.1$

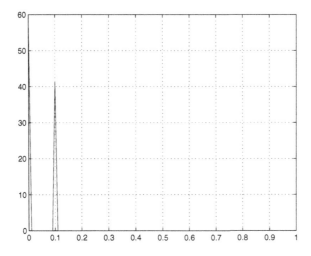

Fig. 5.13 Reconstruction of a 2-point distribution, $p = 0.4, x = 0.1$

A natural idea seems to be to take the arithmetic average. From the resulting curve Fig. 5.18, it is clear that the smoothing is far from perfect. We believe that this imperfection is that the arithmetic average works the best only for normal distributions, and our distributions are different from normal. For such distributions, the

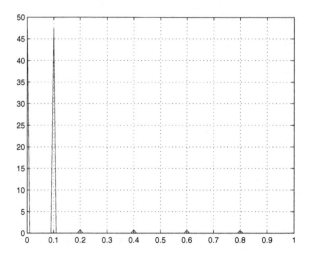

Fig. 5.14 Reconstruction of a 2-point distribution, $p = 0.5$, $x = 0.1$

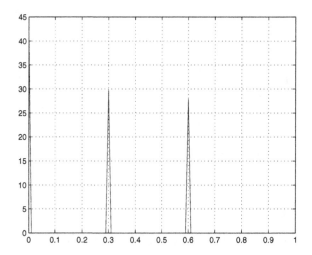

Fig. 5.15 Reconstruction of a 3-point distribution, $p_0 = 0.4$, $p_1 = 0.3$, $x_1 = 0.3$, and $x_2 = 0.6$. This distribution is reconstructed correctly

median provides the most robust averaging; see, e.g., [71]. Because of this, we replace each reconstructed value $\rho(x)$ with the *median* of all the reconstructed values $\rho(x')$ for all values $x' \in [x - \varepsilon, x + \varepsilon]$. After such smoothing, we get a much better reconstruction; see Fig. 5.19.

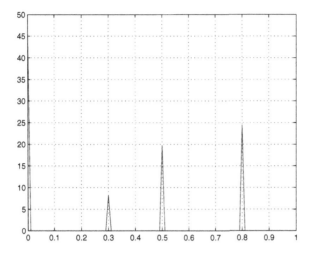

Fig. 5.16 Reconstruction of a 3-point distribution, $p_0 = 0.4$, $p_1 = 0.3$, $x_1 = 0.5$, and $x_2 = 0.8$; reconstruction leads to an additional peak at $x = 0.3$

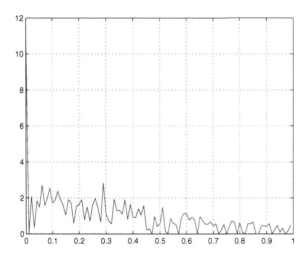

Fig. 5.17 Reconstruction of an asymmetric triangular distribution – before smoothing

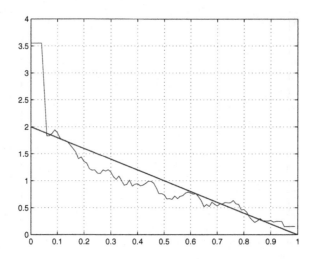

Fig. 5.18 Reconstruction of an asymmetric triangular distribution – after arithmetic-average smoothing

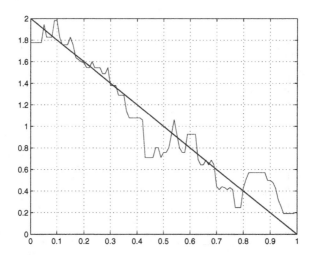

Fig. 5.19 Reconstruction of an asymmetric triangular distribution – after median smoothing

5.8 Images are Easier to Restore than 1-D Signals: A Theoretical Explanation of a Surprising Empirical Phenomenon

The above results lead an theoretical explanation of a surprising empirical phenomenon. Similar techniques are often used to restore 1-D signals and 2-D images from distorted ("blurred") observations.

- From the purely mathematical viewpoint, 1-D signals are simpler, so it should be easier to restore signals than images.
- However, in practice, it is often easier to restore a 2-D image than to restore a 1-D signal.

In this section, we provide a theoretical explanation for this surprising empirical phenomenon.

Empirical Fact. In his presentations at the IEEE World Congress on Computational Intelligence WCCI'2012 (Brisbane, Australia, June 10–15, 2012), J. M. Mendel mentioned a puzzling fact [46]:

- replacing usual type-1 fuzzy techniques [31, 50, 78] with type-2 techniques (see, e.g., [45, 47]) often drastically improves the quality of 2-D *image* processing,
- on the other hand, similar type-2 methods, in general, do not lead to any significant improvement in processing 1-D *signals* (e.g., in predicting time series).

This is not just a weird property of type-2 techniques: J. Mendel recalled that he encountered a similar phenomenon in the 1980s and early 1990s, when he was applying more traditional statistical methods to processing geophysical signals and images; see, e.g., [44].

Why This Is Surprising. From the purely mathematical viewpoint, a 1-D signal means that we have intensity values depending only on one variable (time), while a 2-D image means that we have intensity values depending on two variables – namely, on two spatial coordinates. From this viewpoint, signals are a simplified 1-D version of the 2-D images. It is therefore natural to expect that it is easier to reconstruct a 1-D signal than a 2-D image – but this is not what we observe.

What We Do in This Section. In this section, we provide a theoretical explanation for the above surprisingly empirical phenomenon.

Comment. This justification is an additional argument that a picture is indeed worth a thousand words :-)

General Description of Distortion. Both in signal and in image processing, the observed signal is somewhat distorted ("blurred"):

- for signals, the observed value $\tilde{x}(t)$ at a moment t depends not only on the actual signal $x(t)$ at this moment of time, but also on the values of the signal $x(t')$ at nearby moments of time t';

- similarly, for images, the value $\widetilde{I}(x,y)$ that we observe at a 2-D point with coordinates (x,y) depends not only on the actual intensity $I(x,y)$ at this spatial point, but also on spatial intensities $I(x',y')$ at nearby points (x',y').

Both in signal processing and in image processing, this distortion is usually well-described as *convolution* (see, e.g., [56]), i.e., as a transformation from $x(t)$ to

$$\widetilde{x}(t) = \int K(t-t') \cdot x(t') \, dt' \qquad (5.25)$$

and from $I(x,y)$ to

$$\widetilde{I}(x,y) = \int K(x-x',y-y') \cdot I(x',y') \, dx' \, dy'. \qquad (5.26)$$

Our goal is to reconstruct the original signal $x(t)$ (or the original image $I(x,y)$) from the distorted observations $\widetilde{x}(t)$ (or $\widetilde{I}(x,y)$).

An additional complication is that the functions $K(t)$ (or $K(x,y)$) which describe the distortion are not known exactly.

What We Prove. In precise terms, we prove that when we do not have any information about the distortion function, then, in the ideal no-noise case:

- it is, in general, *not possible* to uniquely reconstruct the original 1-D signal;
- however, it is, in general, *possible* to uniquely reconstruct the original 2-D image.

Comment. In real life, when noise *is* present, the reconstruction is, of course, no longer unique, but the above empirical fact shows that, with some accuracy, reconstruction of 2-D images is still possible.

Convolution Can Be Naturally Described in Terms of Fourier Transform. It is well known (see, e.g., [6, 56, 73]) that formula for the convolution can be simplified if we use Fourier transform

$$\widehat{x}(\omega) = \frac{1}{\sqrt{2\pi}} \cdot \int_{-\infty}^{\infty} x(t) \cdot \exp(i \cdot \omega \cdot t) \, dt, \qquad (5.27)$$

$$\widehat{I}(\omega_x, \omega_y) = \frac{1}{2\pi} \cdot \int \int I(x,y) \cdot \exp(i \cdot (\omega_x \cdot x + \omega_y \cdot y)) \, dx \, dy. \qquad (5.28)$$

Namely, in terms of Fourier transforms, the formulas (5.25) and (5.26) take a simple form

$$\widehat{\widetilde{x}}(\omega) = \widehat{K}(\omega) \cdot \widehat{x}(\omega), \qquad (5.29)$$

$$\widehat{\widetilde{I}}(\omega_x, \omega_y) = \widehat{K}(\omega_x, \omega_y) \cdot \widehat{I}(\omega_x, \omega_y). \qquad (5.30)$$

In Practice, We Only Observe Discrete Signals (Images). In practice, we only observe finitely many intensity values. For a signal, we measure the values $\widetilde{x}_k$ corresponding to moments $t_k = t_0 + k \cdot \Delta$, where Δt is the time interval between two consecutive measurements. For an image, we similarly usually measures intensities $\widetilde{I}_{k,\ell l}$ corresponding to the grid points $(x_k, y_\ell) = (x_0 + k \cdot \Delta x, y_0 + \ell \cdot \Delta y)$.

In the Discrete Case, Fourier Transforms Can Be Reformulated in Terms of Polynomials. Based on the observed discrete value, we cannot recover the original signal (image) with high spatial resolution, we can only hope to recover the values $x_k \stackrel{\text{def}}{=} x(t_k)$ and $I_{k,\ell} \stackrel{\text{def}}{=} I(x_k, y_\ell)$ the original signal (image) on the same grid (or an even sparser set. In terms of the observed and actual grid values, the Fourier transform formulas take the form of an integral sums, such as

$$\widehat{\widetilde{x}}(\omega) = \frac{1}{\sqrt{2\pi}} \cdot \sum_{k=0}^{N} (\widetilde{x}_k \cdot \Delta t) \cdot \exp(i \cdot \omega \cdot k \cdot \Delta t). \tag{5.31}$$

In terms of $s \stackrel{\text{def}}{=} \exp(i \cdot \omega \cdot \Delta t)$, this formula takes the polynomial form $\widehat{\widetilde{x}}(\omega) = P_{\widetilde{x}}(s)$, where

$$P_{\widetilde{x}}(s) \stackrel{\text{def}}{=} \sum_{k=0}^{N} (\widetilde{x}_k \cdot \Delta t) \cdot s^k. \tag{5.32}$$

Similarly, we have $\widehat{x}(\omega) = P_x(s)$ and $\widehat{K}(\omega) = P_K(s)$, where

$$P_x(s) \stackrel{\text{def}}{=} \sum_{k=0}^{N} (x_k \cdot \Delta t) \cdot s^k, \quad P_K(s) \stackrel{\text{def}}{=} \sum_{k=0}^{N} (K_k \cdot \Delta t) \cdot s^k. \tag{5.33}$$

For these polynomials, we have

$$P_{\widetilde{x}}(s) = P_K(s) \cdot P_x(s). \tag{5.34}$$

This equality holds for infinitely many different values $s \stackrel{\text{def}}{=} \exp(i \cdot \omega \cdot \Delta x)$ corresponding to infinitely many different values ω.

It is known that the difference between two polynomials of degree N is also a polynomial of the same degree and thus, this difference can have no more than N roots. So, if the difference between two polynomials is equal to 0 for infinitely many values s, this means that this difference is identically 0, i.e., that the equality (5.34) holds for all possible values s.

Similarly, for the 2-D image case, for $s_x \stackrel{\text{def}}{=} \exp(i \cdot \omega \cdot \Delta x)$ and $s_y \stackrel{\text{def}}{=} \exp(i \cdot \omega \cdot \Delta y)$, we get

$$P_{\widetilde{I}}(s_x, s_y) = P_K(s_x, s_y) \cdot P_I(s_x, s_y), \tag{5.35}$$

where

$$P_{\widetilde{I}}(s_x, s_y) \stackrel{\text{def}}{=} \sum_{k=0}^{N} \sum_{\ell=0}^{N} \left(\widetilde{I}_{k,\ell} \cdot \Delta x \cdot \Delta y \right) \cdot s_x^k \cdot s_y^\ell, \tag{5.36}$$

$$P_K(s_x, s_y) \stackrel{\text{def}}{=} \sum_{k=0}^{N} \sum_{\ell=0}^{N} (K_{k,\ell} \cdot \Delta x \cdot \Delta y) \cdot s_x^k \cdot s_y^\ell, \tag{5.37}$$

$$P_I(s_x, s_y) \overset{\text{def}}{=} \sum_{k=0}^{N} \sum_{\ell=0}^{N} (I_{k,\ell} \cdot \Delta x \cdot \Delta y) \cdot s_x^k \cdot s_y^\ell. \tag{5.38}$$

In Terms of the Resulting Polynomials, Reconstructing a Signal (Image) Means Factoring a Polynomial. In terms of the polynomial equalities (5.34) and (5.35), the problem of reconstructing a signal or an image takes the following form: we know the product of two polynomials, and we want to reconstruct the factors that lead to this product.

In 1-D Case, There Are Many Ways to Represent a Polynomial as a Factor of Two Others. In the 1-D case, as we have mentioned earlier in this chapter, each complex-valued polynomial $P_{\tilde{x}}(s)$ of degree N has, in general, N complex roots $s^{(1)}$, $s^{(2)}$, etc., and can, therefore, be represented as $|P(s)|^2 = \text{const} \cdot (s - s^{(1)}) \cdot (s - s^{(2)}) \cdot \ldots$. In this case, there are many factors, so there are many ways to represent it as a product of two polynomials.

In the 2-D Case, Polynomial Factorization Is almost always Unique. As we shown earlier in this chapter, in contrast to the 1-D case, in which each polynomial can be represented as a product of polynomials of 1st order, in the 2-D case, a generic polynomial *cannot* be represented as a product of polynomials of smaller degrees.

Since a generic 2-D polynomial cannot be factorized, this means that, in general, from the product $P_{\tilde{I}}(s_x, s_y)$ of two 2-variable polynomials (5.35), we can uniquely determine both factors – in particular, we can uniquely determine the polynomial $P_I(s_x, s_y)$.

Based on the the observed value $\tilde{I}(x, y)$, we can determine $P_{\tilde{I}}(s_x, s_y)$, and from the polynomial $P_I(s_x, s_y)$, we can uniquely determine its coefficients $I_{k,\ell} \cdot \Delta x \cdot \Delta y$, and thus, we can determine the original intensity values $I_{k,\ell} = I(x_k, y_\ell)$. So, in the absence of noise, we can indeed (almost always) uniquely reconstruct a 2-D image but not a 1-D signal. The statement is proven.

References

1. Aubinet, M., Vesala, T., Papale, D. (eds.): Eddy Covariance – A Practical Guide to Measurement and Data Analysis. Springer, Heidelberg (2012)
2. Ausiello, G., Crescenzi, P., Kann, V., Marchetti-Spaccamela, A., Protasi, M.: Complexity and Approximation: Combinatorial Optimization Problems and Their Approximability Properties. Springer, Heidelberg (1999)
3. Averill, M.G.: A Lithospheric Investigation of the Southern Rio Grande Rift, University of Texas at El Paso, Department of Geological Sciences, Ph.D. Dissertation (2007)
4. Averill, M.G., Miller, K.C., Keller, G.R., Kreinovich, V., Araiza, R., Starks, S.A.: Using expert knowledge in solving the seismic inverse problem. International Journal of Approximate Reasoning 45(3), 564–587 (2007)
5. Bishop, C.M.: Pattern Recognition and Machine Learning. Springer, New York (2006)
6. Bracewell, R.N.: Fourier Transform and Its Applications. McGraw Hill, New York (1978)
7. Brassard, G., Høyer, P., Tapp, A.: Quantum counting. In: Larsen, K.G., Skyum, S., Winskel, G. (eds.) ICALP 1998. LNCS, vol. 1443, pp. 820–831. Springer, Heidelberg (1998)
8. Candès, E., Romberg, J., Tao, T.: Robust Uncertainty Principles: Exact Signal Reconstruction from Highly Incomplete Frequency Information. IEEE Transactions on Information Theory 52, 489–509 (2006)
9. Cormen, T.H., Leiserson, C.E., Rivest, R.L., Stein, C.: Introduction to Algorithms. MIT Press, Cambridge (2009)
10. Cryer, J.D., Chan, K.-S.: Time Series Analysis. Springer, New York (2010)
11. Elad, M.: Sparse and Redundant Representations. Springer (2010)
12. Fishburn, P.C.: Utility Theory for Decision Making. John Wiley & Sons Inc., New York (1969)
13. Fishburn, P.C.: Nonlinear Preference and Utility Theory. The John Hopkins Press, Baltimore (1988)
14. Garloff, J.: Zur intervallmässigen Durchführung der schnellen Fourier-Transformation. ZAMM 60, 291–292 (1980)
15. Goldsztejn, A.: Private communication (2007)
16. Grover, L.: A fast quantum mechanical algorithm for database search. In: Proc. 28th ACM Symp. on Theory of Computing, pp. 212–219 (1996)
17. Grover, L.K.: Quantum mechanics helps in searching for a needle in a haystack. Phys. Rev. Lett. 79(2), 325–328 (1997)

18. Herrera, J.: A robotic tram system used for understanding the controls of Carbon, water, of energy land-atmosphere exchange at Jornada Experimental Range. Abstracts of the 18th Symposium of the Jornada Basin Long Term Ecological Research Program, Las Cruces, Mexico, July 15 (2010)

19. Hiramatsu, A., Huynh, V.-N., Nakamori, Y.: A behavioral decision model based on fuzzy targets in decision making using weather information. Journal of Advanced Computational Intelligence and Intelligent Informatics 12(5), 435–442 (2008)

20. Hole, J.A.: Nonlinear high-resolution three-dimensional seismic travel time tomography. Journal of Geophysical Research 97, 6553–6562 (1992)

21. Huynh, V.-N., Nakamori, Y.: Behavioral decision analysis using fuzzy targets. In: Proceedings of the Fifth International Conference of the Thailand Econometric Society, Chiang Mai, Thailand, January 12-13 (2012)

22. Huynh, V.-N., Nakamori, Y., Lawry, J.: A probability-based approach to comparison of fuzzy numbers and applications to target oriented decision making. IEEE Transactions on Fuzzy Systems 16(2), 371–387 (2008)

23. Huynh, V.-N., Nakamori, Y., Ryoke, M., Ho, T.B.: Decision making under uncertainty with fuzzy targets. Fuzzy Optimization and Decision Making 6(3), 255–278 (2007)

24. Huynh, V.-N., Yan, H.B., Nakamori, Y.: A target-based decision making approach to consumer-oriented evaluation model for Japanese traditional crafts. IEEE Transactions on Engineering Management 57(4), 575–588 (2010)

25. Jaimes, A.: Net ecosystem exchanges of Carbon, water and energy in creosote vegetation cover in Jornada Experimental Range. Abstracts of the 18th Symposium of the Jornada Basin Long Term Ecological Research Program, Las Cruces, Mexico, July 15 (2010)

26. Jaimes, A., Tweedie, C.E., Peters, D.C., Herrera, J., Cody, R.: GIS-tool to optimize site selection for establishing an eddy covariance and robotic tram system at the Jornada Experimental Range, New Mexico. Abstracts of the 18th Symposium of the Jornada Basin Long Term Ecological Research Program, Las Cruces, New Mexico, July 15 (2010)

27. Jaimes, A., Tweedie, C.E., Peters, D.C., Ramirez, G., Brady, J., Gamon, J., Herrera, J., Gonzalez, L.: A new site for measuring multi-scale land-atmosphere Carbon, water and energy exchange at the Jornada Experimental Range. Abstracts of the 18th Symposium of the Jornada Basin Long Term Ecological Research Program, Las Cruces, Mexico, July 15 (2010)

28. Jaulin, L., Kieffer, M., Didrit, O., Walter, E.: Applied Interval Analysis, with Examples in Parameter and State Estimation, Robust Control and Robotics. Springer, London (2001)

29. Jaynes, E.T.: Probability Theory: The Logic of Science. Cambridge University Press, Cambridge (2003)

30. Keeney, R.L., Raiffa, H.: Decisions with Multiple Objectives. John Wiley and Sons, New York (1976)

31. Klir, G.J., Yuan, B.: Fuzzy Sets and Fuzzy Logic. Prentice Hall, Upper Saddle River (1995)

32. Kohli, R., Krishnamurthi, R., Mirchandani, P.: The minimum satisfiability problem. SIAM Journal on Discrete Mathematics 7(2), 275–283 (1994)

33. Kreinovich, V.: Interval computations and interval-related statistical techniques: tools for estimating uncertainty of the results of data processing and indirect measurements. In: Pavese, F., Forbes, A.B. (eds.) Data Modeling for Metrology and Testing in Measurement Science, pp. 117–145. Birkhauser-Springer, Boston (2009)

34. Kreinovich, V., Lakeyev, A., Rohn, J., Kahl, P.: Computational Complexity and Feasibility of Data Processing and Interval Computations. Kluwer, Dordrecht (1997)

35. Kreinovich, V., Longpré, L.: Fast quantum algorithms for handling probabilistic and interval uncertainty. Mathematical Logic Quarterly 50(4/5), 507–518 (2004)

36. Laney, C., Cody, R., Gallegos, I., Gamon, J., Gandara, A., Gates, A., Gonzalez, L., Herrera, J., Jaimes, A., Kassin, A., Kreinovich, V., Nebesky, O., Pinheiro da Silva, P., Ramirez, G., Salayandia, L., Tweedie, C.: A cyberinfrastructure for integrating data from an eddy covariance tower, robotic tram system for measuring hyperspectral reflectance, and a network of phenostations and phenocams at a Chihuahuan Desert research site. Abstracts of the FLUXNET and Remote Sensing Open Workshop "Towards Upscaling Flux Information from Towers to the Globe, Berkeley, California, June 7-9, p. 48 (2011)

37. Lee, X., Massman, W., Law, B.: Handbook of Micrometeorology – A Guide for Surface Flux Measurements. Springer, Heidelberg (2011)

38. Lees, J.M., Crosson, R.S.: Tomographic inversion for three-dimensional velocity structure at Mount St. Helens using earthquake data. Journal of Geophysical Research 94, 5716–5728 (1989)

39. Liu, G., Kreinovich, V.: Fast convolution and fast Fourier transform under interval and fuzzy uncertainty. Journal of Computer and System Sciences 76(1), 63–76 (2010)

40. Longpré, L., Servin, C., Kreinovich, V.: Quantum computation techniques for gauging reliability of interval and fuzzy data. International Journal of General Systems 40(1), 99–109 (2011)

41. Luce, R.D., Raiffa, R.: Games and Decisions: Introduction and Critical Survey. Dover, New York (1989)

42. Maceira, M., Taylor, S.R., Ammon, C.J., Yang, X., Velasco, A.A.: High-resolution Rayleigh wave slowness tomography of Central Asia. Journal of Geophysical Research 110(B6) (2005)

43. Martinez, M., Longpré, L., Kreinovich, V., Starks, S.A., Nguyen, H.T.: Fast quantum algorithms for handling probabilistic, interval, and fuzzy uncertainty. In: Proceedings of the 22nd International Conference of the North American Fuzzy Information Processing Society, NAFIPS 2003, Chicago, Illinois, July 24-26, pp. 395–400 (2003)

44. Mendel, J.M.: Optimal Seismic Deconvolution: an Estimation-Based Approach. Academic Press, New York (1983)

45. Mendel, J.M.: Uncertain Rule-Based Fuzzy Logic Systems: Introduction and New Directions. Prentice-Hall, Upper Saddle River (2001)

46. Mendel, J.M.: Proceedings of the IEEE World Congress on Computational Intelligence, WCCI 2012, Brisbane, Australia, June 10-15 (2012)

47. Mendel, J.M., Wu, D.: Perceptual Computing: Aiding People in Making Subjective Judgments. IEEE Press and Wiley, Piscataway, New Jersey (2010)

48. Moncrieff, J.B., Malhi, Y., Leuning, R.: The propagation of errors in long-term measurements of land-atmospheric fluxes of carbon and water. Global Change Biology 2, 231–240 (1996)

49. Moore, R.E., Kearfott, R.B., Cloud, M.J.: Introduction to Interval Analysis. SIAM Press, Philadelphia (2009)

50. Nguyen, H.T., Walker, E.A.: First Course In Fuzzy Logic. CRC Press, Boca Raton (2006)

51. Nielsen, M.A., Chuang, I.L.: Quantum Computation and Quantum Information. Cambridge University Press, Cambridge (2000)

52. Ochoa, O.: Towards a fast practical alternative to joint inversion of multiple datasets: model fusion. Abstracts of the 2009 Annual Conference of the Computing Alliance of Hispanic-Serving Institutions, CAHSI, Mountain View, California, January 15-18 (2009)

53. Ochoa, O., Velasco, A.A., Kreinovich, V., Servin, C.: Model fusion: a fast, practical alternative towards joint inversion of multiple datasets. Abstracts of the Annual Fall Meeting of the American Geophysical Union AGU 2008, San Francisco, California, December 15-19 (2008)

54. Ochoa, O., Velasco, A., Servin, C.: Towards Model Fusion in Geophysics: How to Estimate Accuracy of Different Models. Journal of Uncertain Systems 7(3), 190–197 (2013)
55. Ochoa, O., Velasco, A.A., Servin, C., Kreinovich, V.: Model Fusion under Probabilistic and Interval Uncertainty, with Application to Earth Sciences. International Journal of Reliability and Safety 6(1-3), 167–187 (2012)
56. Orfanidis, S.: Introduction to Signal Processing. Prentice Hall, Upper Saddle River (1995)
57. Papadimitriou, C.H.: Computational Complexity. Addison Wesley, San Diego (1994)
58. Rabinovich, S.: Measurement Errors and Uncertainties: Theory and Practice. American Institute of Physics, New York (2005)
59. Raiffa, H.: Decision Analysis. Addison-Wesley, Reading (1970)
60. Ramirez, G.: Quality data in light sensor network in Jornada Experimental Range. Abstracts of the 18th Symposium of the Jornada Basin Long Term Ecological Research Program, Las Cruces, New Mexico, July 15 (2010)
61. Richardson, A.D., et al.: Uncertainty quanitication. In: Aubinet, M., Vesala, T., Papale, D. (eds.) Eddy Covariance – A Practical Guide to Measurement and Data Analysis, pp. 173–209. Springer, Heidelberg (2012)
62. Rockafeller, R.T.: Convex Analysis. Princeton University Press, Princeton (1970)
63. Sanchez, R., Argaez, M., Guillen, P.: Sparse Representation via l^1-minimization for Underdetermined Systems in Classification of Tumors with Gene Expression Data. In: Proceedings of the IEEE 33rd Annual International Conference of the Engineering in Medicine and Biology Society EMBC 2011 "Integrating Technology and Medicine for a Healthier Tomorrow", Boston, Massachusetts, August 30-September 3, pp. 3362–3366 (2011)
64. Sanchez, R., Servin, C., Argaez, M.: Sparse fuzzy techniques improve machine learning. In: Proceedings of the Joint World Congress of the International Fuzzy Systems Association and Annual Conference of the North American Fuzzy Information Processing Society, IFSA/NAFIPS 2013, Edmonton, Canada, June 24-28, pp. 531–535 (2013)
65. Servin, C., Ceberio, M., Jaimes, A., Tweedie, C., Kreinovich, V.: How to describe and propagate uncertainty when processing time series: metrological and computational challenges, with potential applications to environmental studies. In: Pedrycz, W., Chen, S.-M. (eds.) Time Series Analysis, Model. & Applications. ISRL, vol. 47, pp. 279–299. Springer, Heidelberg (2013)
66. Servin, C., Huyhn, V.-N., Nakamori, Y.: Semi-heuristic target-based fuzzy decision procedures: towards a new interval justification. In: Proceedings of the Annual Conference of the North American Fuzzy Information Processing Society, NAFIPS 2012, Berkeley, California, August 6-8 (2012)
67. Servin, C., Jaimes, A., Tweedie, C., Velasco, A., Ochoa, O., Kreinovich, V.: How to Gauge Accuracy of Measurements and of Expert Estimates: Beyond Normal Distributions. In: Proceedings of 3rd World Conference on Soft Computing, San Antonio, Texas, December 15-18 (2013)
68. Servin, C., Kreinovich, V.: Images are easier to restore than 1-D signals: a theoretical explanation of a surprising empirical phenomenon. Journal of Uncertain Systems 8 (to appear, 2014)
69. Servin, C., Ochoa, O., Velasco, A.A.: Probabilistic and interval uncertainty of the results of data fusion, with application to geosciences. Abstracts of 13th International Symposium on Scientific Computing, Computer Arithmetic, and Verified Numerical Computations, SCAN 2008, El Paso, Texas, September 29-October 3, p. 128 (2008)

70. Servin, C., Tweedie, C., Velasco, A.: Towards a more realistic treatment of uncertainty in Earth and environmental sciences: beyond a simplified subdivision into interval and random components. Abstracts of the 15th GAMM-IMACS International Symposium on Scientific Computing, Computer Arithmetic, and Verified Numerical Computation, SCAN 2012, Novosibirsk, Russia, September 23-29, pp. 164–165 (2012)

71. Sheskin, D.J.: Handbook of Parametric and Nonparametric Statistical Procedures. Chapman and Hall/CRC Press, Boca Raton (2011)

72. Shumway, R.H., Stoffer, D.S.: Time Series Analysis and Its Applications. Springer, New York (2010)

73. Sneddon, I.N.: Fourier Transforms. Dover Publ., New York (2010)

74. Statnikov, A., Tsamardinos, I., Dosbayev, Y., Aliferis, C.: GEMS: A system for automated cancer diagnosis and biomarker discovery from microarray gene expression data. International Journal of Medical Informatics 74, 491–503 (2005)

75. Vapnik, V.: The Nature of Statistical Learning Theory, 2nd edn. Springer, New York (2000)

76. Yan, H.-B., Huynh, V.-N., Nakamori, Y.: Target-oriented decision analysis with different target preferences. In: Torra, V., Narukawa, Y., Inuiguchi, M. (eds.) MDAI 2009. LNCS, vol. 5861, pp. 182–193. Springer, Heidelberg (2009)

77. Yan, H.-B., Huynh, V.-N., Nakamori, Y.: A group nonadditive multiattribute consumer-oriented Kansei evaluation model with an application to traditional crafts. Annals of Operations Research 195(1), 325–354 (2012)

78. Zadeh, L.A.: Fuzzy sets. Information and Control 8, 338–353 (1965)

Index

Printed in the United States
By Bookmasters